Affordable Housing for Smart Villages

This book initiates a fresh discussion of affordability in rural housing set in the context of the rapidly shifting balance between rural and urban populations. It conceptualises affordability in rural housing along a spectrum that is interlaced with cultural and social values integral to rural livelihoods at both personal and community level. Developed around four intersecting themes: explaining houses and housing in rural settings; exploring affordability in the context of aspirations and vulnerability; rural development agendas involving housing and communities; and construction for resilience in rural communities, the book provides an overview of some of the little understood and sometimes counter-intuitive best practices on rural affordability and affordable housing that have emerged in developing economies over the last thirty years. Drawing on practice-based evidence this book presents innovative ideas for harnessing rural potential, and empowering rural communities with added affordability and progressive development in the context of housing and improved living standards.

- For a student aspiring to work in rural areas in developing countries it is an introduction to and map of some key solutions around the critical area of affordable housing
- For the rural development professional, it provides a map of a territory they rarely see because they are absorbed in a particular rural area or project
- For the academic looking to expand their activities into rural areas, especially in rural housing, it provides a handy introduction to a body of knowledge serving 47% of the world's population, and how this differs from urban practice
- For the policy makers, it provides a map for understanding the dynamics around rural affordability, growth potential and community aspirations helping them to devise appropriate intervention programs on rural housing and development

Hemanta Doloi is a senior lecturer in Construction Management discipline and Project Director of the Smart Villages project at the Faculty of Architecture, Building and Planning of the University of Melbourne, Australia. He is the founding director of the Smart Villages Lab (SVL) and the lead author of *Planning, Housing and Infrastructure for Smart Villages*.

Sally Donovan is a research fellow with the Smart Villages Lab in the Faculty of Architecture, Building and Planning at the University of Melbourne, Australia. Dr Donovan has over ten year's experience researching environmental management and environmental policy development. She is a co-author of the book *Planning, Housing and Infrastructure for Smart Villages*.

Affordable Housing for Smart Villages

Hemanta Doloi and Sally Donovan

Routledge
Taylor & Francis Group

LONDON AND NEW YORK

First published 2020
by Routledge
2 Park Square, Milton Park, Abingdon, Oxon OX14 4RN

and by Routledge
52 Vanderbilt Avenue, New York, NY 10017

Routledge is an imprint of the Taylor & Francis Group, an informa business

© 2020 Hemanta Doloi and Sally Donovan

British Library Cataloguing-in-Publication Data
A catalogue record for this book is available from the British Library

Library of Congress Cataloging-in-Publication Data
Names: Doloi, Hemanta Kumar, author. | Donovan, Sally, 1953- author.
Title: Affordable housing for smart villages / Hemanta Doloi and Sally Donovan.
Description: k, Abingdon, Oxon ; New York : Routledge, 2020. | Includes bibliographical references and index.
Identifiers: LCCN 2019027779 (print) | LCCN 2019027780 (ebook) | ISBN 9780367190774 (hbk) | ISBN 9780367190781 (pbk) | ISBN 9780429200250 (ebk) | ISBN 9780429577963 (adobe pdf) | ISBN 9780429573743 (mobi) | ISBN 9780429575853 (epub)
Subjects: LCSH: Smart cities. | Dwellings--Design and construction--Cost effectiveness. | Housing--Planning. | Community development. | Rural development.
Classification: LCC TD159.4 .D649 2020 (print) | LCC TD159.4 (ebook) | DDC 333.33/8091734--dc23
LC record available at https://lccn.loc.gov/2019027779
LC ebook record available at https://lccn.loc.gov/2019027780

ISBN: 978-0-367-19077-4 (hbk)
ISBN: 978-0-367-19078-1 (pbk)
ISBN: 978-0-429-20025-0 (ebk)

Typeset in Goudy
by Taylor & Francis Books
Printed and bound by CPI Group (UK) Ltd, Croydon, CR0 4YY

Contents

List of figures   vi
List of tables   viii
List of boxes   ix
Preface   x
Acknowledgements   xii

1 Introduction   1

2 Housing in rural settings   6

3 The nature–culture determinants of rural housing   24

4 Affordable houses   48

5 Housing affordability   68

6 Materials and resources in construction of affordable houses   87

7 Global practices in rural development   109

8 Vulnerability in rural communities   133

9 Resilience in rural communities   150

10 Sustained growth and development   166

11 Epilogue   171

Index   173

Figures

1.1	Factors associated with affordable housing in Smart Villages	2
2.1	Five key dimensions of affordable housing in rural settings	6
2.2	Typical size of houses and distance from urban centres	18
2.3	Neighbourhood communication and density of houses	19
3.1	Nature–culture–critical regionalism of rural housing design	26
3.2	Elements of functional requirements of a rural house	26
3.3	Extreme weather conditions and degree of tolerance	30
3.4	Key features of vernacular architecture in hot and humid climate	31
3.5	Tea garden bungalow in India	33
3.6	A typical Mishing house on stilt foundation in a flood prone area in Assam	35
3.7	A typical Karbi house in rural Assam	37
3.8	A typical Japanese house with local materials	40
3.9	Vernacular architecture of a typical Japanese double storey house	41
3.10	Sliding door in Japanese house	42
4.1	Defining affordability	51
4.2	Owning versus renting a house	55
4.3	Legislative support on affordability solutions	61
5.1	Location and configuration of rural housing	69
6.1	Construction of affordable houses	88
6.2	Plan of the two square metre shed	97
6.3	Complete LGS frame erected in less than four hours	98
6.4	Concrete-less steel plate and micro-pile assembly	98
6.5	Sectional view of the LGS framed housing model	99
6.6	LGS framed finished housing model on slab foundation	99
6.7	LGS framed finished housing model on stilt foundation	100
7.1	PMAYG house of Resident P.1	111
7.2	PMAYG house of Resident P.2	112
7.3	PMAYG house of Resident P.3	112
7.4	PMAYG house of Resident P.4	113
7.5	PMAYG house of Resident P.5	114

7.6	PMAYG house of Resident P.6	115
7.7	PMAYG house of Resident P.7	115
7.8	PMAYG house of Resident P.8	116
7.9	PMAYG house of Resident P.9	117
7.10	Self-made house of Resident S.1	118
7.11	Self-made house of Resident S.2	119
7.12	Self-made house of Resident S.3	119
7.13	Self-made house of Resident S.4	120
7.14	Self-made house of Resident S.5	121
7.15	Self-made house of Resident S.6	122
7.16	Self-made house of Resident S.7	122
7.17	Self-made house of Resident S.8	123
7.18	Self-made house of Resident S.9	124
7.19	Self-made house of Resident S.10	125
7.20	Typical Assamese Kacha house	129
7.21	Vernacular Assam Type house	130
8.1	Vulnerability associated with sub-standard housing	134
9.1	Development of housing and resilient community	152
9.2	Vernacular stilt house	158
9.3	Assam type house	159
10.1	Framework for developing strategy and policy on affordable houses	167

Table

7.1   Results of survey questions from Assam, India   126

Boxes

Case study P.1: PMAYG housing scheme beneficiary, Assam,
India (Resident P.1)   110
Case study P.2: PMAYG housing scheme beneficiary, Assam,
India (Resident P.2)   111
Case study P.3: PMAYG housing scheme beneficiary, Assam,
India (Resident P.3)   112
Case study P.4: PMAYG housing scheme beneficiary, Assam,
India (Resident P.4)   113
Case study P.5: PMAYG housing scheme beneficiary, Assam,
India (Resident P.5)   113
Case study P.6: PMAYG housing scheme beneficiary, Assam,
India (Resident P.6)   114
Case study P.7: PMAYG housing scheme beneficiary, Assam,
India (Resident P.7)   115
Case study P.8: PMAYG housing scheme beneficiary, Assam,
India (Resident P.8)   116
Case study P.9: PMAYG housing scheme beneficiary, Assam,
India (Resident P.9)   117
Case study S.1: self-made home, Assam, India (Resident S.1)   118
Case study S.2: self-made home, Assam, India (Resident S.2)   118
Case study S.3: self-made home, Assam, India (Resident S.3)   119
Case study S.4: self-made home, Assam, India (Resident S.4)   120
Case study S.5: self-made home, Assam, India (Resident S.5)   121
Case study S.6: self-made home, Assam, India (Resident S.6)   121
Case study S.7: self-made home, Assam, India (Resident S.7)   122
Case study S.8: self-made home, Assam, India (Resident S.8)   123
Case study S.9: self-made home, Assam, India (Resident S.9)   124
Case study S.10: self-made home, Assam, India (Resident S.10)   124

Preface

The premise of this book is that the underlying assumption of previous affordable housing schemes that rural communities have limited or no affordability, is fundamentally flawed. A completely new approach is necessary to accurately assess community affordability, understand potentials and design affordable housing that has the potential to not just meet but to exceed community expectations by upgrading their living standards. Based on the analysis of survey data collected from 37 rural villages (spanning over 2000 households) in Assam, India, it is evident that there is a clear disparity between public housing schemes and public aspirations for affordable housing. Such research-based evidence forms a solid basis for stimulating an educated argument on the core topic of affordability and affordable housing in this book.

Currently, in the Smart Villages Lab at the University of Melbourne, the authors are undertaking empirical research to develop new knowledge and theories with a particular focus on rural planning, housing, infrastructure and governance. The aim of the research is to create new knowledge and support public policies focusing on eradicating poverty in emerging economies such as India. Rapid development programmes that target rural population require an in-depth investigation of public policies and their underlying governance mechanisms that result in effective delivery of affordable housing and associated services. Over the past few decades, many grand schemes to develop affordable housing for rural development have failed to produce expected outcomes. Traditional top down housing solutions, laden with public policies have increased the gap between rural and urban communities across numerous fronts. The Smart Villages research, initiated from a project funded by the State Government of Assam, is centred on fundamental improvements in rural infrastructure, local skills and knowledge required for rural development. Researching into this area of critical need, the authors co-authored a first book *Planning, Housing and Infrastructure for Smart Villages*, Routledge, UK (2019).

Housing and infrastructure provision is one of the key areas of disparity between urban and rural communities. The Smart Villages lab's research provides a unique opportunity for improving rural housing policy by incorporating culture and community at the core of decision-making. By undertaking the fieldwork in this research, a series of intellectual enquiries have been initiated based on accurate representation of the individual households with a comprehensive set of socio-economic data. These enquiries provided a solid basis for formulating new policies on affordable housing from the perspective of the house owners. Highlighting the underlying knowledge and processes around creation of community housing in a global context, the book provides a good basis for comparing and contrasting the opinions of rural residents with expert viewpoints, bringing an insightful debate on affordability and affordable rural housing that will appeal to a global audience.

Acknowledgements

The Government of Assam being the sponsor of the Smart Villages project at the University of Melbourne, the authors sincerely express their gratitude to a number of senior officials including honourable ministers within the government for creating the needs and providing the opportunity for fundamental research in the Smart Villages programme. At the outset, the authors wish to acknowledge the visionary leadership of the past Chief Minister of Assam, Mr Tarun Gogoi who took a very personal interest in creation of the collaboration between the State of Assam and the University of Melbourne. In the early part of the project, the assistance received from many senior leaders in the government including Mr Paban Borthakur, Mr Rajiv Bora, Mr Ranjit Hazarika is highly valuable. The past Chief Secretary Mr Vinod Pipersenia was highly instrumental in placing the Assam Engineering College as the technical agency to work in the collaboration.

Dr Atul Bora who is a past Director of Technical Education (DTE) and the current principal of Assam Engineering College, supported every aspect of the collaborative endeavours between the two institutions. His insightfulness, guidance, support and leadership in the Smart Villages project is highly commendable.

The authors sincerely appreciate the Honourable Chief Minister of the Government of Assam, Shri Sarbananda Sonowal and Honourable Minister for Finance Dr Himanta Biswa Sarma for their kind support, valuable guidance and encouragement in continued collaboration in the project. The generous financial support given by the State Government of Assam in the project enabled the authors to pull together credible resources of international calibre to support and develop shared capacities in the project. Special thanks go to the Honourable Minister for Education Mr Siddhartha Bhattacharya for extending his inspiration and support in the Smart Villages project. The authors are also grateful to Shri Maninder Singh for his valuable support and advice during his visit to Melbourne.

In regard to this book, the authors are thankful to many people who tirelessly supported the project from its inception to completion. Special thanks go to Dr Kiran Shinde whose research assistance in collecting some early literature for conceptualising the idea for this book is phenomenal. The authors are also grateful for the advice and guidance provided by the acquisition editor and the entire editorial team of the publisher in guiding us through the lifecycle of the project. Among many people, two masters students from Jorhat Engineering College, Assam namely Mrikangka Dutta Barua and Pranamee Barua deserve special mention for supporting gathering of some useful field data for the book. Prof Vijay from IIT Delhi along with a few of his colleagues in the Centre for Rural Development and Technology were highly instrumental in providing valuable insights during an interview with the first author on affordability of housing in an Indian context.

Last but not least, the authors are truly appreciative of the support and leadership of Prof Julie Willis, Dean; Associate Prof Andrew Hutson, Deputy Dean and Rebecca Bond, Executive Director of the Faculty of Architecture, Building and Planning at the University of Melbourne. The space and facilities provided to the Smart Villages Lab (SVL) within the faculty is one of the key requirements for being productive in research and producing high quality output such as this book within a short period of time.

1 Introduction

Rural villages were once synonymous with self-sufficient agricultural communities. Lack of electricity and telecommunications links saw these communities essentially cut off from the rest of world. The local architecture in these areas reflected this self-sufficiency and isolation as houses were built by members of the local community, from locally sourced materials, with design features that had evolved in response to the local climate and culture. In more recent decades, the industrialisation of agriculture has created larger farms with fewer employees as many labour jobs are replaced with technology. These production processes are very cost-effective meaning these farms can charge lower prices for their produce. Thus, small-holder farms are unable to compete in the market place and can no longer generate sufficient income from agricultural activities. Similarly, the number of labour positions available on larger farms are declining. Waves of young able-bodied citizens are migrating to urban areas in search of better opportunities leaving rural communities with high proportions of elderly residents and small-holder farms growing food for their own sustenance but not generating any income. While rural communities are still cut-off due to poor infrastructure they are becoming more dependent on external resources for survival. The increase in access to media has also influenced once isolated rural communities by providing a window into how other people live. This has altered their expectations, leading to a preference for "urban" lifestyles.

The concept of a "Smart Village" has evolved in response to this changing nature of rural areas. This would see the transformation of these struggling communities into thriving, self-sufficient ones. At the heart of the Smart Village concept is the creation of income generating opportunities, to help attract young people back to these areas, coupled with the enhancement of local skill level to reduce their dependence on external resources. For example, providing a microgrid for the generation of electricity would bring the benefits of electrification without the risk of blackouts accompanied by connection to the national grid. Training members of the local community to perform repairs and maintenance on the system would ensure that it remains in good working order without the need for external support. Increased electrification would go hand in hand with improved telecommunications access, bringing rural villages a world of information and enhancing income generation potential. Small businesses will gain information that could help enhance the quality and quantity of their production. For example, farmers could learn ways to improve their crop yields. The internet also provides isolated communities with a way to source better supplies and increase their customer base. However, perhaps the most important part of enhancing people's lives is providing them with an affordable, comfortable, functional and safe place to live in.

For most people worldwide, the cost of accommodation takes up the greatest proportion of their income. Due to the vastly unequal distribution of incomes, people with vastly different financial resources are competing in the same housing market. As a result, poorer people are often forced to either live in appalling conditions or move to afford their accommodation. This destroys the socio-economic diversity of communities creating concentrated pockets of cheap low-quality housing. These areas are undesirable both as a place to live and a place to visit. Thus, it will be difficult for businesses to thrive, meaning they are more likely to move their premises to a different area. This has the doubly negative impact of moving jobs away from potential employees and moving services away from residents in areas with poor housing (Whitehead, 2007).

Providing appropriate affordable housing in diverse communities will be a key component of transforming decrepit rural areas into Smart Villages. To do this, the provision of housing must go beyond the basic concept of a "roof over the head" to embrace a range of features that could lead to the improvement of livelihoods. These concepts are summarised in Figure 1.1.

Figure 1.1 Factors associated with affordable housing in Smart Villages

As can be seen, costs and affordability only make up one of the factors for consideration. It is also important to consider:

Location and local settings, understanding whether the house is located in close proximity to essential services, employment opportunities.

The culture, value and family setup are also important. The use of a house goes beyond simply a place to sleep, also providing a place for children to play and study, a work space for small businesses, and a place for entertaining guests. Households with strong religious beliefs often also require spatial layouts that allow them to practise rituals.

The local climatic conditions should also be factored into the house design. Vernacular architecture in most regions has evolved in response to the local climate (Zhai and Previtali, 2010). Thus, housing in warmer climates retains coolness during the daytime and cools down quickly in the evening, while in cooler climates housing is designed to retain heat. Modern architecture often fails to capture this climatic responsiveness meaning either the interior space will be uncomfortable or that households will be required to spend money on heating and cooling their internal space.

One of the most significant costs of rural housing is the construction and therefore, reducing costs is important for achieving affordability. Reducing construction costs is often considered a reduction in quality, however, there are ways to avoid this compromise. Relying on local resources in terms of both building materials and labour could help make housing more cost-effective, while also increasing local employment opportunities.

In many cultures a home is a place for entertaining guests and thus it is important that the aesthetic appearance of the house is something to be proud of. Many people feel their house is an important status symbol and will have direct bearing the way they are perceived by other members of their community. Ensuring a pleasing aesthetic while still using low-cost materials is an important challenge.

Poorly built houses are not just unpleasant but can also be dangerous to live in. People have the right to feel safe in their own home, therefore, reducing costs should not compromise on quality.

Perhaps the most important part of making a house affordable is ensuring longevity. While it may be possible to build a relatively cheap, aesthetically pleasing house, it cannot be considered affordable if it is severely damaged during storms or other disasters. This is particularly the case for regions where such occurrences are frequent. There some evidence that the intensity of natural disasters is increasing due to climate change, and that this scary trend is projected to continue, therefore ensuring houses are resilient to any natural disasters projected to impact a region is also an essential consideration.

This book will describe the importance of incorporating all these factors into affordable house designs, and ways that these can be achieved. It draws on evidence from previous affordable housing schemes and their successes and failures in the following structure.

Rural communities have typically been associated with agriculture and traditional lifestyles. However, in the face of an increasingly globalised world and increasingly industrialised agriculture the nature of rural communities is changing. Chapter 2 describes some of the new types of rural communities that are emerging in the modern world.

The changing face of rural communities is visible in the alteration of rural architecture. Traditional vernacular housing designs typified by locally sourced building materials are being replaced with modern brick and concrete houses. While there are some benefits to modernisation, vernacular designs evolved in response to the location's climate and culture and so have some features that are more appropriate. Their use of locally sourced renewable building materials also makes them superior in the emerging need to embrace sustainability into the construction of buildings. In Chapter 3 we explore some of the design features of vernacular structures that are superior to modern designs due to their climatic responsiveness and cultural significance.

Nearly every region in the world appears to be experiencing housing shortages these days. Thus, people everywhere are struggling to afford accommodation. As a result, many are either finding themselves cutting back on other basic essentials to cover their accommodation costs, living in squalid or overcrowded conditions, or living in an inconvenient location. Countries are often judged by the state of their housing; thus, governments are struggling to respond to this ever-increasing problem. In Chapter 4 we look at different definitions of affordable housing and discuss the importance of going beyond monetary considerations to incorporate adequacy and availability. We then look at a range of specific affordable housing initiatives incorporating both government and non-governmental case studies, and their successes and failures.

The house in itself is not the only factor affecting affordability. The location of the house is also an important consideration. Convenient access to employment opportunities and essential services will impact on a family's quality of life and can either be hindered or enhanced by the location of their accommodation. For many families a house is not just a place to sleep in but is a place to perform many daily activities including work, study and socialising. The ability to perform all these activities will be affected by the layout of the house and its surrounding landscape, impacting both cost of living and quality of life. In Chapter 5 we explore different ways affordability can be impacted by the location and layout of a house.

The obvious solution to the chronic global housing shortage is to construct more houses. While this sounds simple enough, there are two challenges to achieving this. Firstly, finding suitable land to build on and granting citizens tenure over this land. Secondly, the actual construction, which entails the costs of building materials and labour. Both of these can be difficult to access in rural communities, increasing their costs compared to urban areas. Reducing construction costs is often considered synonymous with compromising on quality. However, by embracing vernacular design features and using a mixture of traditional and modern building materials cost-efficient quality housing could be achieved. In Chapter 6 we describe in detail different issues relating to land tenure and present a range of construction options that could significantly decrease the overall costs of housing, without compromising on quality.

As part of the research for this book the authors visited a remote rural region of India and met with locals to find out about their housing and their feelings about their homes. Some of these people were living in houses that had been built by the government under the "housing for all" initiative, while others were self-built. Chapter 7 summarises the results of this field study along with discussion around the suitability of government-funded housing provision.

One criticism of vernacular houses is the perception that they are flimsy and will collapse easily in extreme conditions. However, others believe the reverse to be true. Vernacular housing was designed to cope with local conditions, including the natural disasters that frequent some regions. In Chapter 8 we look at the impact that some of the worst disasters in recent history have had on housing in rural communities, including earthquakes, tsunamis, cyclones, floods, droughts, fires and conflict.

Architects and engineers have developed design features specifically to increase the resilience of housing to these natural disasters. Some vernacular features have also been designed in response to natural disasters, in areas where these are a frequent occurrence. In Chapter 9 we describe some of these specific features pertaining to the different types of disasters that occur frequently in some regions. We compare modern and vernacular features in determining which of these is most effective at ensuring resilience while still ensuring house construction is affordable for rural citizens. Thus, we look at ways of making rural housing that is both resilient and affordable.

References

Whitehead, C. M. E. (2007). Planning policies and affordable housing: England as a successful case study? *Housing Studies*, 22(1), 25–44.

Zhai, Z. J., and Previtali, J. M. (2010). Ancient vernacular architecture: Characteristics categorization and energy performance evaluation. *Energy and Buildings*, 42, 357–365.

2 Housing in rural settings

Rural areas are perceived as places with a strong community spirit where all the residents know each other and look out for each other; as agrarian based economies where generations of the same family have lived and worked the same land for millennia. This idealised scenario of rural life resonates globally, in both developed and developing regions. Within this universal picture of rural life, every rural community has distinct culture and associated traditions. These cultural traits derive from a combination of influences from the national level, the state or other subdivision level and at the village level. Even within a village, individual households may have different cultural practices. They may come from religious beliefs, ethnicity or a combination of these factors. In the state of Assam, India there are over 100 different ethnic groups occupying rural areas with associated cultural practices. Different ethnicities and religions coexist in rural villages the world over from the Islamic and Zoroastrian peoples of Iran, to the Sinhalese and Tamil communities of Sri Lanka (Haigh et al., 2016).

Figure 2.1 shows five key dimensions of affordable housing in rural settings. The elements are livelihood, culture, climate, external influence, fitness for purpose and safety. While all these elements are important to be intact, in practice, there is a compromise in adaptive cultural shifts in rural communities.

Figure 2.1 Five key dimensions of affordable housing in rural settings

The different cultural traditions that come from all levels of society are steeped in history, and it is easy to become sentimental when a tradition is lost. In reality though, the traditions of rural communities everywhere have been evolving over time. The loss of traditions can be due to many factors, sometimes these are practical reasons. For example, in some regions, when the head of household became too old to work the land, the land was divided up evenly between all the male children of the family, and the extended family would work together with other members of the community to build a house for each of the children's families on their land. The elderly parents would live on with one of the children's families. Of course, over time this practice became unfeasible as the parcels of land became smaller and smaller until they were no longer viable for agricultural activities. In other cases cultural practices are influenced by historical events. Periods of conflict for example can influence land ownership and housing design as is the case in Saudi Arabia. Here many rural houses have look out posts and few windows on the exterior of their house, opting instead for an internal courtyard due to the long term conflict experienced in this region (Saleh, 2000).

In more recent times, as the impact of globalisation has pervaded even the most remote of rural communities, changes have been occurring in a more ubiquitous way. One of the biggest impacts is the industrialisation of agriculture. This has seen small farms amalgamated into a single large farm producing more output with fewer employees. Thus, small-holder farming struggles to be profitable and has largely become an exercise in self-sufficiency. The option of low-skilled workers finding a job as farm labourers is also dwindling due to the intensification of production rates and automation of processes. Finding alternative forms of employment in rural areas is extremely difficult for low-skilled workers, driving young people, especially men to migrate to urban areas.

The other major impact is the result of the pervasion of television and internet access into rural areas. This is altering lifestyles in various ways. Johnson (2001) for example, describes the impact television has had on rural India. The behavioural changes described range from small alterations to daily routines as people begin to organise their daily activities to fit in around their favourite shows to influencing aspirations. Seeing the way other people live raises awareness of the different standards of living experienced throughout the world and drives desires to have a lifestyle emulating those seen on TV. While there are some positive results of this, for example, Johnson noted that families had developed better communication and incidents of domestic violence decreased, it is also turning people away from traditional rural lifestyles and driving urban migration.

Urban migration does not only affect immediate families. Every time a person moves away from a rural community that is one less potential customer for any businesses in the area, and one less citizen requiring services, making the provision of services less and less efficient (Kilkenny, 2010). Thus, the decline of agriculture in rural areas negatively affects the whole community in many ways. It also means the average age of rural residents is increasing, many elderly residents living in close-knit communities with few young people to assist them with their needs. Thus, while the efficiency of providing essential services, especially healthcare is decreasing, the necessity is increasing.

Conversely, urbanites on high incomes are tiring of the harsh realities of urban life such as pollution, overcrowding and high crime rates and thus desire a more rural lifestyle. They dream of living in a bucolic area surrounded by the beautiful scenery of nature and being friends with their neighbours. These people are in a conflicting situation as they are expecting an idealised version of life in a rural village, yet at the same time are used to the conveniences provided by living in an urban area. They may use their high socio-economic status to attempt to alter the rural communities they move into to fit in with their ideals.

All of these influences are combining to change the face of rural lifestyles. In this chapter we examine four types of emerging rural communities. The first three build on the classifications described by Bański and Wesołowska (2010) in Poland. The first classification we define as *commuter communities* – rural villages located in close proximity to an urban centre; the second is *tourism communities* – rural areas located near areas of outstanding natural beauty; thirdly we consider the remaining *agrarian communities*, that is, areas that lack the attractions that can bring urbanites into a region and thus remain as largely agricultural areas; and finally we look at *manufactured communities*, which can either be communities built up around a single major employer (also referred to as dormitory towns (Beer, 1998)) or newly established areas that are supposed to develop self-sufficiency through the housing and infrastructure provided by the developers. We draw on examples from both developed and developing economies to look at the impact of these changes on the concept of the rural idyll, and ways these changes can be regulated to ensure they retain the benefits of rural life.

Commuter communities

Rural communities that are located close to urban areas can attract commuters, who are willing to spend a longer period of time travelling to and from work each day, to obtain the perceived benefits of living in a rural village. These include less pollution, peaceful and friendly atmosphere, better safety, bigger house and access to nature. There are also potential benefits these commuters could bring to rural areas, for example, increasing the efficiency of services and increasing the customer base for local businesses. Unfortunately, an increased commuter base can also have negative consequences for rural communities. Namely, the increased demand for housing in these areas leads to an increase in housing prices. The commuters, who are earning the higher salaries of urban based jobs, can afford to pay more for their accommodation. As a result, they will price local residents who work in the village, usually on lower incomes, out of the housing market (Ziebarth, Prochaska-Cue, and

Shrewsbury, 1997; Cook et al., 2002). The opportunities for rental accommodation in rural communities is often limited, as most housing is owner-occupied (Morton, Allen, and Li, 2004). Through our survey of residents of Majuli Island, Assam, India, we found that 1.2 per cent of households were renting out of 2300 households across 37 villages. Thus, it is unlikely low-income earners will be able to continue living in the area.

Another consequence of the increased demand for housing will be pressure to increase the local housing stock. This will mean either increasing housing density or developing on green land. Thus, either housing sizes will decrease, or green spaces will be diminished, two of the assets that attracted commuters to pastoral life in the first place. The decision of whether to increase density or develop on green space is a very divisive one for many rural communities, creating conflict between those expecting to profit from the new development and those expecting to lose out from it. Options for increasing the housing density could be to tear down existing properties and replace with newer more compact housing such as apartment complexes or smaller single-family homes. In the coastal town where one of the authors lives in Australia, the traditional large single-family home properties have largely been replaced with two to four two storey units that have limited garden sizes. People are willing to forgo a private garden to live in close proximity to the beach.

Local residents in these communities often oppose this type of development as traditional building styles can be strongly associated with the character of a particular village. Thus, as traditional buildings are replaced with more modern ones, locals may feel their town is losing its character. Another option is to maintain the façade of existing buildings, thus preserving local character, and to renovate the inside to turn a large traditional single-family home into a block of apartments. While this option may be preferable to locals, it can be a lot more expensive and may limit the overall number of households that can be developed.

Developing on green space is equally controversial for rural communities. For example, struggling small farms may be purchased by developers and subdivided. On the one hand, this be could considered buying out a failing business and turning the land into profit. On the other, it could be seen as turning pleasant green spaces into blocks of housing. The use of forest land or other natural areas is even more controversial, as it could lead to loss of biodiversity. There are various schools of thought in relation to the importance of biodiversity conservation. Many people believe that all species inherently have the right to exist. Others consider it crucial to protect the ecosystem services we as humans rely on for survival that could be lost through the destruction of the natural environment. Development of only a part of a natural area can disturb the whole ecosystem leading to large scale negative consequences.

On a local scale, some members of rural villages may rely on ecosystem services for their income. In India, forest dwelling communities harvest the leaves of the palash tree and weave them into plates. These provide a sustainable alternative to disposable plastic plates (Chavan et al., 2016). In Southern Mixteca, Mexico pine resin is extracted to make turpentine (Hernandez-Aguilar et al., 2017). Thus, development of the forest area will take away their livelihoods.

In developed countries such as the U.K. the development of green spaces often receives strong public opposition. Sustainability principles are usually cited as the main concern of local residents who argue that increasing housing density in already urbanised areas has a lower environmental impact than developing on green spaces. However, it has been asserted that while sustainability is used to block many housing developments, environmental impact assessments are not always conducted to confirm this. Coombes (2009) investigated the use of environmental issues to block new housing developments and found that the true motivation of local residents was fears that their property values would be negatively impacted by the new development. Locals were also concerned over the impact a new development would have on the atmosphere of the community and the aesthetic appeal of the area. Where a housing development is specifically targeting low-income earners the opposition is exacerbated. Bramley and Watkins (2009) also explored this issue in the U.K. and found that housing developments in rural areas that do obtain planning permission are generally aimed at higher income earners. We will look at these negative perceptions of affordable housing more fully in Chapter 4.

Rural areas surrounding urban centres in China are attracting not only residential property development, but industrial and commercial facilities are also looking for more affordable locations for their premises here as well (Tan and Li, 2013). This competition for land has led to fears over food shortages as more and more agricultural land is developed. In response, the government has passed a regulation preventing the loss of arable land. The regulation requires that the total area of arable land remains constant. The responsibility for ensuring the land area is retained falls to the local government level, thus for local authorities in areas surrounding major urban centres the only option is to increase density of development. Detached single family homes will have to make way for multistorey apartment blocks. This prospect is certainly not attractive to many commuter residents, whose reasons for moving to a rural area included having a larger home, and more outdoor space. However, realistically many areas may have to face the fact that there is no alternative solution. Where development of green space is exhausted, increasing housing density of already developed spaces is the only option for catering to the exponentially increasing population.

In the U.S. a compromised approach was attempted which allowed development of dense pockets of housing, while large areas of green space remained protected (Ryan, 2006). The commuter community in this instance sacrificed the desire to have a big house with a large garden in favour of being able to live in the vicinity of a forested area where they could enjoy outdoor activities on their doorstep. The developers built the housing estates in such a way that they were not visible from any of the major roads, ensuring the region retained its characteristic rural charm. In this instance, tourism was one of the major industries of the region, so retention of rural character was essential to the continued success of the local businesses. The scheme proved to be successful as the commuting residents were satisfied with having access to the forest, while the long-term communities' strong tourism industry was not harmed.

Another component of rural village life is having communal areas with locally owned and run shops and markets, and other sorts of businesses as well as community centres where local residents can gather to socialise. One of the perceived benefits of commuter communities is that retailers in these villages will see an increase in business due to the increased population. In reality, large retail chain stores will also see the opportunity in the area and will attempt to purchase a large tract of land nearby to build huge retail parks (Good, 2002). The lower prices these stores can offer will attract consumers away from the small retail businesses located in the town centre, resulting in them being driven out of business. Residents working locally in the service industry will subsequently end up working in these parks, further away from their home, often for a lower salary, with the added burden of a longer commute. Thus, the thriving town centre ideal of rural life will be lost. Many of the stores will be converted to residential properties to fulfil the increased demand for housing and the communal meeting places will disappear.

In Ireland, architects attempted to create a housing development with a rural village atmosphere by developing high density neighbourhoods with communal spaces (Donovan and Gkartzios, 2014). However, the project was largely a failure as the residents did not use the communal spaces in the way intended. The authors argue that community spirit is not a product that can be manufactured but is something that evolves naturally over time as residents become acquainted. Instead their neighbourhoods had the typical suburban atmosphere of neighbours that never speak to each other.

The state of Massachusetts in the U.S. is renowned for its rural charm, while also being home to major urban centres. The attraction to settle in these rural villages has seen the advent of malls and the subsequent loss of the small-town character as local retail stores are put out of business in many regions (Good, 2002). Communal meeting spaces that were once the main hub of these small communities have been lost as people meet in the mall food courts and other places which are largely only accessible by car. Many of the downtown business premises and other community spaces have been converted to housing to meet the increasing demand. To prevent this loss of rural character, a few of the small towns in this state have attempted to implement policies to better control their development. Good (2002) describes the approach used by one of these towns that has successfully retained its characteristic charm while also allowing significant development.

The first step is to identify what constitutes the rural charm, that is, what specific parts of the town need to be preserved to retain the town's character. This may include anything that is meaningful to local residents including historic buildings or parks. The next step is to implement a policy protecting these features from development. Where new development is to take place, design guidelines are established so that the style of the new housing development is in keeping with the character of the town. Every stage of the process attempts to involve as many members of the community as possible through education campaigns and public forums ensuring awareness of development proposals and planning regulations is widespread. During these discussions different members of the township are often at odds. Local developers see the potential profit in converting property to residences to address the increased housing demand and subsequent increase in prices potential; while locals who are fond of local history consider preservation of historical features to be the highest priority. These people need to be brought together with council representatives to ensure everyone's opinions are heard and considered seriously, and the ultimate solution should provide a compromise. Other important components are encouraging local residents to support local businesses. While many locals may say they want to retain the local high street feel of the small town, they may do their own shopping at the out of town strip malls. Encouraging people to utilise local businesses to ensure they can compete with the large retail chains is essential to their survival.

Thus, while an increased commuter population has led to the detriment of many once rural communities, Good (2002) shows that with careful planning and management it is possible to retain the best of both worlds. Applying these development principles could lead to retention of the safe, close knit atmosphere of rural villages, while still allowing the population to increase.

Tourism communities

The second type of rural community is emerging in the vicinity of areas of outstanding natural beauty or sites of special scientific interest which attract tourists. As struggling agricultural households sell up in these regions, properties targeting the burgeoning tourism industry will take their place. These properties mainly cater to second homes, purchased by urbanites for weekend or vacation use, retirees, or short-term rentals such guest houses, bed and breakfasts and so on. The increased tourism industry will create jobs in the local area, both directly through providing tours, working in guest houses, and indirectly by increasing trade for local shops and restaurants. The tourism trade will provide alternative employment for low-skilled workers who would previously have found employment in agriculture.

On the downside, the increased demand for properties in the area will drive up house prices meaning the locals, who are working in low-income service industry jobs, will be competed out of the local housing market.

Options for renting may also be hindered as property owners may prefer using their properties as short-term lets for tourists. There are many advantages to short-term leasing over long-term, which will be discussed in more detail in Chapter 4.

For residents that own a traditional farm house, there is the potential to turn their property into a guest house. Retaining their farm may add to the tourist attraction of the property, while allowing the homeowners to diversify their income sources giving them greater financial security.

An example of a rural area that has transitioned from an agricultural to a tourism-based region is in the Annapurna Massif of Nepal (Nepal, 2007). The region contains around 30 high peaks, including one over 8000 metres, which attract many mountain climbers and trekkers. The fact that the region is widely inhabited gives trekkers a unique cultural experience on top of the beautiful views afforded from the many trekking routes (Wikipedia, 2019). Many residents have successfully converted their traditional farming houses into guest houses. In fact, Nepal (2007) notes that many residents who had moved away in search of better employment opportunities have been able to return to their family home to run it as a guest house. While more modern tourist accommodation is around, many tourists are interested in learning about local culture when they visit a new area and are attracted to the idea of staying in a vernacular building. Also, the ever-increasing volume of tourists visiting the region means smaller, less well-developed guest houses are still finding customers. As they generate profits they may be able to make improvements to their homes, especially adding a bathroom where this is lacking as this can be a major turn off for tourists.

The impact of tourism on local culture is another area of concern. Some people consider that local culture can be enhanced by engaging tourists into traditions. Classes teaching crafts, dancing and cooking are all examples of popular tourist activities. Thus, the community can also take advantage of increased tourism to turn their local traditions into a small business. There is a risk however, that these cultural experiences can be commandeered by non-local tour operators, who want to offer their perception of a cultural experience, rather than an authentic one. In doing this they will either compromise the local community into altering their culture, or bring in outsiders to operate these courses, denying locals this income generating opportunity. The concept of tourism can also compromise cultural traditions, for example, in the Cook Islands, etiquette dictates that hospitality must be reciprocated (Berno, 1999). Opening your home to guests for monetary compensation is thus a break with tradition. However, a research study into the impact of tourism in the area found that locals had managed to adapt by distinguishing between locals and outsiders in the way they treat guests. Any native Cook Islanders were welcomed as a guest, with the understanding that if the home owner visited the guest's home they would be equally welcome, while visitors from outside the Cook Islands were offered accommodation for monetary compensation.

While an influx of tourists can bring in profits for small businesses, tourism is often seasonal. Therefore, as permanent residents of rural communities decline it may become difficult for small businesses to survive. It is also notable that while the decline of small-holder agriculture is one of the main reasons for the alteration of rural communities, in some regions, despite their lack of profitability small-holder farms can form part of the tourist attraction. Therefore, it can be in the interest of other local businesses to support the preservation of these traditional agricultural properties (Shucksmith and Rønningen, 2011). Traditional agricultural practices are returning to favour as awareness of the detrimental impacts of industrial agriculture is growing. Therefore, preserving the knowledge of traditional agricultural techniques is also of interest. Shucksmith and Rønningen (2011) noted that this situation is applicable in Norway and Scotland and considered ways to make small-holder farms viable in these regions. At the moment government subsidies are offered to eligible farms but due to the unstable nature of politics relying on this is risky. The possibility of charging tourists for farm tours was floated. While certainly viable, concerns over whether this could disrupt the running of the farm were raised. Niche commodity markets, for example, offering organic certified produce, allow farmers to sell produce at a premium rate and can offer a way for small farms to compete with the large industrial agriculture. The government subsidies could be better used to bring these farms up to code for niche commodity status, giving them a longer-term economic situation.

In general, large scale unregulated development runs the risk of destroying the natural features that attracted tourists to the region in the first place. This has been seen in Spain, the Caribbean Islands (Bickford et al., 2017) and Poland (Bański and Wesołowska, 2010). Bickford et al., (2017) state that tourism development must incorporate regulatory tools including *corporate social responsibility* (CSR) and *social licence to operate* (SLO) to ensure any development adheres to sustainability principles, preserving environmental features and local culture, while enhancing the local economy. As stated above tourism development creates jobs. While it is assumed these will go to locals, this has not always been the case. In Zimbabwe, tourism companies brought in more experienced staff from elsewhere to work in their hotels and resorts only employing locals in the lowest-paid part time jobs (Chirenje, 2017). The development did little to improve the economic situation of the local community. CSR and SLO must also be used to ensure companies are required to employ locals, provide them with adequate training allowing them to attain higher positions, and pay them fair wages.

Thus, while increased tourism can bring jobs or other income generating opportunities to declining rural communities, it can also bring increased house prices, over development and the loss of local culture. Developing a tourist economy in a way that ensures local residents can afford to remain in their home should be a priority. As we have seen in Nepal it is possible for the transition to be successful, where not only did current locals manage to enhance their livelihoods, but relatives who had been forced to move away for work were able to return home (Nepal, 2007). However, in other regions the transition has been less successful, for example in the U.K. where second home buyers priced locals out of the housing market forcing them to move away (Rye, 2011) or Spain and the Caribbean where increased development led to the destruction of the natural beauty that provided the tourist attraction (Bickford et al., 2017). A compromise such as the cluster housing proposed by Ryan (2006) shows that it is possible to increase the availability of affordable housing without damaging the natural beauty of an area as long as the development is managed by good policies.

Agrarian communities

Agrarian-based rural communities still exist in regions that are not close to urban centres and have no areas of particular attraction to tourists. As the produce sold in local markets is dominated by industrial agriculture, small-holder farms in these regions are largely operating for self-sufficiency. In developing regions, these isolated agrarian communities can lack access to basic amenities such as electricity and telecommunications equipment. Thus, their agricultural activities require time-consuming manual techniques, leaving residents little time to pursue other forms of employment. This coupled with a lack of employment opportunities in these areas is driving migration to urban areas in search of better opportunities. These people are often unable to migrate as families, due to their poverty status, rather men migrate alone leaving women and children behind to fend for themselves.

Thus, the demographic distribution of these types of rural areas is changing to higher proportions of elderly people, and more single-parent households. While the men may be making money in the urban regions, passing this on to their families is hindered as many of these regions are not served by banks. In the meantime, women must provide for their families.

Thus, the role of women in agrarian communities is also changing. In many cultures, women traditionally would have filled their days with household duties and caring for other members of the family, such as children and elderly relatives. Women whose husbands have migrated for work, now need to fill the role of both parents. This will involve spending long hours in income generating activities, agricultural activities or both. Even where both members of a married couple are able to remain in the village, it is likely that they will both need to work and work longer hours to cover the increasing cost of living. There are benefits arising from this situation, as women are attaining greater independence and reduced exposure to domestic violence.

The role of children will also need to adapt due to the changing nature of rural life. They may no longer be able to devote their time to educational attainment, instead taking on a different role. For girls this usually means taking over household duties, and potentially caring for younger siblings, for boys helping with agricultural activities. Even if children are only required to assist during harvest time, missing some school will mean they fall behind the other students and many find it difficult to catch up and end up dropping out of school altogether. Education is perhaps the most important way to break children out of the poverty cycle, so this loss of opportunity can be detrimental. In South America, an alternative approach to education in rural areas known as Nuevas Escuales, has successfully helped many children (Ferero-Pineda, Escobar-Rodriguez, and Molina, 2006). In this schooling system, children are given a book to work through at their own pace, while the teacher floats around the classroom offering assistance when needed. This way, children can take time off during a harvest, and pick up where they left off, rather than returning to a class that has advanced in their absence. It also means they can work through their studies at home. The programme has been running for long enough so that researchers were able to interview adults who had gone through the programme. They found them more proactive in their community compared to people who had gone through traditional school systems.

Thus far the situation for agrarian communities is coming across as fairly bleak; however, thanks to United Nations Millennium and Sustainable Development Goals, things are slowly improving. Internet access is increasing, mainly through increased availability of smart phones and inroads are slowly being made into electricity access thanks to initiatives such as providing microgrid electricity from renewable resources. Electricity access will allow many of the time-consuming manual labour tasks to be replaced with automated processes, freeing up time for people to pursue other activities. In Nepal for example the electrification of rural villages allowed automation of the rice milling process, which was found to free up 155 hours per year for women and 85 hour per year for men (Legros, Rijal, and Seyedi, 2011).

As we saw in the introduction there are positive and negative impacts of increased access to television and the internet. There are some ways these can specifically help improve agrarian livelihoods. A very simple one is access to weather reports. In developing economies many farmers have no irrigation facilities and instead rely on rainfall to water their crops. Access to information on rainfall can help farmers to know the best time to plant crops, apply pesticides and fertilisers; therefore, access to this information is crucial to improving their overall crop yields.

Information about the use of pesticides and fertilisers could also lead to big improvements, as these are often misused. Many farmers rely on information from either the government or local pesticide and fertiliser sellers (Doloi, Green, and Donovan, 2019). Government officials do not necessarily have any farming experience and will be passing on information they receive from elsewhere, often pesticide and fertilisers manufacturers. As the manufacturers and sellers of the products have a vested interest in boosting sales, their advice often tries to convince people to apply far more than necessary. Excessive application is not only a waste of money, it can also have negative side effects.

For example, excess pesticide use poses a health risk to the farmer during application, can kill insects that are beneficial to plant production, and increases the risk of pesticide resistance. Excess fertiliser will be dispersed into the wider environment where it can cause eutrophication in waterways, and increase cumulation of N_2O, a potent greenhouse gas with 265–298 times the global warming potential of CO_2. In general, internet access would mean access to information on farming techniques that maximise crop yields. Trowbridge (2005) notes that over the years most governments have poured countless amounts of funding into farming subsidies, yet farmers continue to struggle for survival. This funding was often used to purchase pesticides, fertilisers and farming equipment. Instead, if this money had funded training programmes teaching best practice the impact could have been far more effective.

Another opportunity for small-holder farms, as mentioned in the previous section, is niche commodities. These can not only provide a way for small farms to compete with larger farms in the marketplace but can create a tourist attraction potentially opening up more opportunities for agrarian communities to improve their development.

In an increasingly globalised world, it is assumed that everyone needs to be functioning as a part of the global community. However, there are some people who envision self-sufficient agrarian rural communities that are not just subsisting but thriving. Razak, Malik, and Saeed (2013) describes a SMART village initiative targeting rural communities in Malaysia. In the area, older residents (ages 40 plus) are still largely engaged in agriculture, however, younger residents see limited potential for employment in agriculture and wish to pursue other opportunities. The scheme aims to teach this despondent younger generation improved agricultural techniques and to show them the potential for establishing a thriving agricultural community. This way they hope to make rural lifestyles more attractive to young people.

Manufactured communities

Another way to revitalise struggling rural communities is to establish a large industry in the area. This can either be in the form of a large manufacturing plant, a power station or where resources are available a mining community. Jones and Tonts (2003) reports that in the past these schemes have rarely resulted in the desired outcome of seeing a struggling rural community thrive. The main problem is that the locals lack the qualifications and experience for many of the positions and fill only the lowest-paid jobs, while outsiders are brought in to fill the higher-level positions. Due to the limited availability of accommodation in the town, the higher income earners price the locals out of existing accommodation. As a result, they end up living in mobile homes, establishing a new community on the outskirts of the town. Another risk with these types of towns is that if the company goes broke, the whole town will be destroyed.

Similarly, as technology becomes more advanced many manual labour tasks are being replaced by robotics. In India it was found that while these industrial plants initially had a lot of employment opportunities, over time production processes became automated and despite continually increased production levels, the number of jobs actually decreased (Walker, 2008). This created another problem in India where resources such as electricity and water are in limited supply. The addition of a large manufacturing plant requiring huge amounts of both of these saw resources diverted to the industrial plants and away from residential areas, further disadvantaging the local community.

The size of houses, locations and type of neighborhood play significant roles for understanding community characteristics and behaviour in rural settings.

Figure 2.2 shows the typical size of houses over distance from the urban centres among different communities. As seen, the lower income community have smaller houses, and the higher income group have bigger houses closer to the city centre. Their preference for a bigger house is due to convenience and status within the community.

Figure 2.3 depicts communication over density of the housing among the various community types discussed above. In struggling rural regions with very low housing density, it has been suggested that moving the remaining residents into a single larger village could be a more efficient way to provide services and create income generating opportunities (Kilkenny, 2010). In this example, the consideration is dwindling rural communities where the decreased population makes it difficult to economically justify the provision of services. Providing these people with an adequate affordable house in a location closer to services would bring many potential advantages. The potential difficulty in realising this vision is the willingness of residents to move to the new location. Forcing people to move against their will, or refusing to provide services to people unless they move could be controversial. Trowbridge (2005) similarly describes an approach to developing thriving self-sufficient villages for Africa. The developments would provide affordable houses, good infrastructure and adequate services. They would

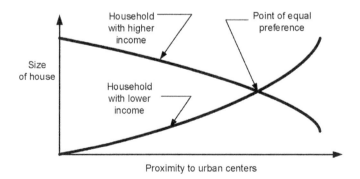

Figure 2.2 Typical size of houses and distance from urban centres

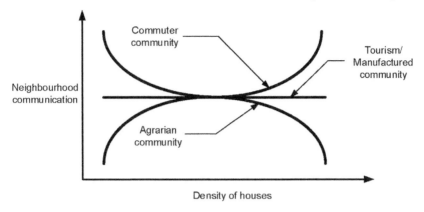

Figure 2.3 Neighbourhood communication and density of houses

also ensure local employment or income generating opportunities to avoid creating concentrated pockets of poverty. The number of residents would be limited to 50,000 to ensure a village atmosphere was created. The intention of this development is to make life in these villages attractive so that people would want to move there rather than forcing people to move.

Creating these amalgamated towns will involve either increasing the housing stock of an existing town or generating a new town from scratch. This causes similar concerns to those mentioned earlier regarding the failed attempt to create a community atmosphere in a commuter community in Ireland, as to whether it is really possible to create such a thing (Donovan and Gkartzios, 2014). Rural communities are often populated by the same family for generations, and the atmosphere may be the result of these long-standing relationships, rather than merely the result of living in close proximity. Thus, forcing groups of disconnected people to live together will not necessarily create a community atmosphere. A potentially effective way to avoid this is engaging people from the early planning stage to make them a part of the scheme. Both the intended beneficiaries of the scheme and existing residents where outsiders are to be moved into an existing town should be consulted throughout the development. Incorporating their input will give them a sense of ownership of the scheme and make it more likely to be successful.

Policy Implications

In this chapter we have explored the way rural communities are being altered by industrialisation and globalisation. In particular, we have considered the impact on disadvantaged members of the community. Low-income households are particularly vulnerable to losing their place in rural villages. Although there are many programmes around aiming to bring development improvements to these people, a lack of interest from other members of the local community can make it difficult for these programmes to have an effect. For example, in India, while religious traditions are often looked on favourably as preserving a higher level of morality, the caste system tradition is being used by local elites in rural areas to justify corrupt behaviour (Tanabe, 2007). Embezzling money from the government, oppressing poor families in the community, and overlooking the needs of the majority to ensure their friends and family are advantaged were all cited as problems in rural India, justified by the traditional caste system. Although this system has officially been abolished it is retained in practice in many areas. Wealthier families feel no desire to see the poorer members of the rural community have more, but rather prefer them to be kept in their place. One of these programmes is legislation requiring that local government committees must include representatives from all caste groups, and this is giving lower castes a voice. However, the higher caste people are not happy about it:

> it is all very well to say that there should be one representative from each hamlet. But that would not work. Committee members must possess a certain quality. What quality do they have? Each family has had its role in the village. We have been the leaders of the village from the past and this quality cannot be acquired overnight.
>
> (quoted in Tanabe, 2007, 564)

The higher caste member quoted above even said that the lower caste person should have been afraid to even suggest such a thing in front of his higher caste superiors. Similar issues with wealthy elites believing that low-income families are inherently inferior have been noted in sub-Saharan African nations and to a certain extent exist in many developed nations as well. While many people understand the benefits greater equality would bring to society, they are not necessarily willing to sacrifice their own wealth to see that equality achieved.

The trend of urban migration largely impacts young people leading to rural communities with increasingly high proportions of elderly residents. In many developing economies, traditionally parents would live with one of their children, usually either the youngest or the eldest male depending on the cultural traditions of the region. Thus, they were guaranteed care in their old age. Now, as many young people have moved away, elderly people are left to fend for themselves. The altered expectations of women are also impacting elderly care. The concept of a woman moving into her husband's parent's house once she is married is often romanticised as being a nice close-knit family living together and supporting each other. In reality the picture is very different. These women are often abused by the husband's family, especially their mother-in-law, and their husbands will rarely side with them against his own parents. In Mexico Pauli (2008) described how women were not allowed to spend time with their own parents and siblings and were cut off from their friends. They were treated as slaves by their in-laws, often subjected to physical abuse. Similar cases have been reported in other locations. The increased awareness of other lifestyles is inspiring these women to escape from this situation. Often their husbands will migrate to the U.S. for work, where they will be earning enough money to build their own home, allowing their wives to have independence. The husbands are happy to build a house for their wife and children because it is a symbol of their wealth and gives them higher status. This desire supersedes any responsibility they feel towards caring for the parents. Thus, alternative forms of care for elderly residents need to be developed.

Varma, Kusuma, and Babu (2009) explores the potential benefits of providing aged care facilities in rural communities of India. They do this by assessing 10 well-being indicators of elderly residents, some living in an aged care facility and some remaining in the village. Of those remaining in the village some were retired, while others, due to their poverty status found it necessary to continue working. Those who remained working had the overall lowest well-being rating of all the people assessed. The findings indicated that those members living in the aged care facility were better off in eight out of the 10 categories assessed, than those remaining in the community. The two exceptions for those remaining in the community were slightly better physical and emotional health. Despite elderly residents in aged care facilities being happier overall, many residents are reluctant to move there. The main reason is due to a stigma around the fact that moving into a retirement community is breaking with tradition in this region. Over time however, it is likely to become essential for many elderly residents due to the changing nature of rural communities.

The phenomena of elderly rural communities are not limited to developing economies. The problem has also been noted in Finland (Rönkkö et al., 2017) and Canada (Ryser and Halseth, 2012). Elderly people are more likely to experience health problems, which can limit them in various ways. They may lose their ability to drive, which can restrict the ability to socialise due to poor public transport links in rural areas. The biggest challenge faced in developed regions is the provision of health care services. Elderly people are more vulnerable and prone to illness and are therefore more likely to need assistance.

One problem with the management of rural areas is the fact that development is considered in isolated pockets. As Trowbridge (2005, 2) says "to a great or lesser extent, all small towns are colonies of the metropolitan areas, relying on them for the majority of their goods and services, and increasingly on charity, but rarely for investments." This connection between rural and urban communities should flow both ways, as rural areas provide food, some industry services and tourist attractions. Therefore, as Rönkkö et al. (2017) proposes, planning policies should reflect the intrinsic link between urban and rural areas by taking a more holistic approach. Instead of treating urban and rural areas as separate entities, the whole region should be considered as an interlinking system with resources flowing both ways.

References

Bański, Jerzy, and Wesołowska, Monika. (2010). Transformations in housing construction in rural areas of Poland's Lublin region – Influence on the spatial settlement structure and landscape aesthetics. *Landscape and Urban Planning*, 94, 116–126.

Beer, Andrew. (1998). Overcrowding, quality and affordability: Critical issues in non-metropolitan rental housing. *Rural Society*, 8, 5–15.

Berno, Tracy. (1999). When a guest is a guest: Cook Islanders view tourism. *Annals of Tourism Research*, 26, 656–675.

Bickford, Nate, Smith, Lindsey, Bickford, Sonja, Bice, Matthew R., and Ranglack, Dustin H. (2017). Evaluating the role of CSR and SLO in ecotourism: Collaboration for economic and environmental sustainability of Arctic resources. *Resources*, 6(2), article 21.

Bramley, Glen, and Watkins, David. (2009). Affordability and supply: The rural dimension. *Planning Practice and Research*, 24, 185–210.

Chavan, S. B., Uthappa, A. R., Sridhar, K. B., Keerthika, A., Handa, A. K., Newaj, Ram, Kumar, Naresh, Kumar, Dhiraj, and Chaturvedi, O. P. (2016). Trees for life: Creating sustainable livelihood in Bundelkhand region of central India. *Current Science*, 111, 994–1002.

Chirenje, Leonard Itayi. (2017). Contribution of ecotourism to poverty alleviation in Nyanga, Zimbabwe. *Chinese Journal of Population Resources and Environment*, 15, 87–92.

Cook, Christine C., Crull, Sue R., Fletcher, Cynthia N., Hinnant-Bernard, Thessalenuere, and Peterson, Jennifer. (2002). Meeting family housing needs: Experiences of rural women in the midst of welfare reform. *Journal of Family and Economic Issues*, 23, 285–316.

Coombes, Mike. (2009). English rural housing market policy: Some inconvenient truths? *Planning Practice and Research*, 24, 211–231.

Doloi, Hemanta, Green, Ray, and Donovan, Sally. (2019). *Planning, housing and infrastructure for Smart Villages*. Abingdon, U.K.: Routledge.

Donovan, Kevin, and Gkartzios, Menelaos. (2014). Architecture and rural planning: "Claiming the vernacular". *Land Use Policy*, 41, 334–343.

Ferero-Pineda, C., Escobar-Rodriguez, D., and Molina, D. (2006). Escuela Nueva's impact on the peaceful social interaction of children in Colombia. In A. W. Little (ed.), *Education for all and multigrade teaching*. Dordrecht, The Netherlands: Springer.

Good, Karen. (2002). *Preservation of small town character in the town center of Rutland, Massachusetts*. Amherst: University of Massachusetts.

Haigh, Richard, Hettige, Siri, Sakalasuriya, Maheshika, Vickneswaran, G., and Weerasena, Lasantha Namal. (2016). A study of housing reconstruction and social cohesion among conflict and tsunami affected communities in Sri Lanka. *Disaster Prevention and Management: An International Journal*, 25, 565–580.

Hernandez-Aguilar, Jose Antonio, Cortina-Villar, Hector Sergoi, Garcia-Barrios, Luis Enrique, and Castillo-Santiago, Miguel Angel. (2017). Factors limiting formation of community forestry enterprises in the Southern Mixteca region of Oaxaca, Mexico. *Environmental Management*, 59, 490–504.

Johnson, Kirk. (2001). Media and social change: The modernizing influences of television in rural India. *Media Culture and Society*, 23, 147–169.

Jones, Roy, and Tonts, Matthew. (2003). Transition and diversity in rural housing provision: The case of Narrogin, Western Australia. *Australian Geographer*, 34, 47–59.

Kilkenny, Maureen. (2010). Urban/regional economics and rural development. *Journal of Regional Science*, 50, 449–470.

Legros, G., Rijal, K., and Seyedi, B. (2011). *Decentralized energy access and the millennium development goals: An analysis of the development benefits of micro-hydropower in rural Nepal*. Rugby, U.K.: Publishing P. A.

Morton, Lois Wright, Lundy Allen, Beverlyn, and Li, Tianyu. (2004). Rural housing adequacy and civic structure. *Sociological Inquiry*, 74, 464–491.

Nepal, Sanjay K. (2007). Tourism and rural settlements: Nepal's Annapurna region. *Annals of Tourism Research*, 34, 855–875.

Pauli, Julia. (2008). A house of one's own: Gender, migration, and residence in rural Mexico. *American Ethnologist*, 35, 171–187.

Razak, Norizan Abudl, Malik, Jalaluddin Abdul, and Saeed, Murad. (2013). A development of smart village implementation plan for agriculture: A pioneer project in Malaysia. In 4th International Conference on Computing and Informatics (ICOCI). Sarawak, Malaysia.

Rönkkö, Emilia, Luusua, Anna, Aarrevaara, Eeva, Herneoja, Aulikki, and Muilu, Toivo. (2017). New resource-wise planning strategies for smart urban-rural development in Finland. *Systems*, 5, 12.

Ryan, Robert L. (2006). Comparing the attitudes of local residents, planners and developers about preserving rural character in New England. *Landscape and Urban Planning*, 75, 5–22.

Rye, Johan Fredrik. (2011). Conflicts and contestations. Rural populations' perspectives on the second homes phenomenon. *Journal of Rural Studies*, 27, 263–274.

Ryser, Laura, and Halseth, Greg. (2012). Resolving mobility constraints impeding rural seniors' access to regionalised services. *Journal of Aging and Social Policy*, 24, 328–344.

Saleh, Mohammed Abdullah Eben. (2000). The architectural form and landscape as a harmonic entity in the vernacular settlements of southwestern Saudi Arabia. *Habitat International*, 24, 455–473.

Shucksmith, Mark, and Rønningen, Katrina. (2011). The uplands after neoliberalism? – The role of the small farm in rural sustainability. *Journal of Rural Studies*, 27, 275–287.

Tan, Minghong, and Li, Xiubin. (2013). The changing settlements in rural areas under urban pressure in China: Patterns, driving forces and policy implications. *Landscape and Urban Planning*, 120, 170–177.

Tanabe, Akio. (2007). Toward vernacular democracy: Moral society and post-postcolonial transformation in rural Orissa, India. *American Ethnologist*, 34, 558–574.

Trowbridge, A. V. (2005). "New towns" – The S.M.A.R.T. alternative to city slums: Lessons to be learned from Soweto to Cosmo City. In *XXXIII World Congress on Housing: Transforming housing environments through design*, edited by International Association for Housing Science Congress (ed.). Pretoria, South Africa.

Varma, G. R., Kusuma, Y. S., and Babu, B. V. (2009). Health-related quality of life of elderly living in the rural community and homes for the elderly in a district of India. *Zeitschrift für Gerontologie und Geriatrie*, 43, 259–263.

Walker, Kathy Le Mons. (2008). Neoliberalism on the ground in rural India: Predatory growth, agrarian crisis, internal colonization, and the intensifictaion of class struggle. *The Journal of Peasant Studies*, 35, 557–620.

Wikipedia. (2019). Annapurna Massif. Wikipedia. Retrieved from https://en.wikipedia.org/wiki/Annapurna_Massif.

Ziebarth, Ann, Prochaska-Cue, Kathleen, and Shrewsbury, Bonnie. (1997). Growth and locational impacts for housing in small communities. *Rural Sociology*, 62, 111–125.

3   The nature–culture determinants of rural housing

Traditionally, people in rural areas designed and built their own housing. Many of these communities were essentially cut off from the rest of the world and thus, their architectural style developed in response to their local environment, local customs and availability of building materials. People were forced to use local resources, both in terms of materials and labour. Construction methods were handed down from generation to generation by word of mouth. Members of the local community worked together to build houses using readily available, renewable resources. The design of houses evolved to incorporate features that suited the local climate and environment, through trial and error, with no knowledge of the scientific reasoning behind these. These houses, referred to as vernacular houses, are often found to be able to maintain comfortable indoor temperatures without the need for mechanical heating and cooling appliances, even in extreme climates, while simultaneously providing a functional layout that caters for the cultural habits of the community.

In Chapter 2 we discussed the impact globalisation has had on rural communities. Rural architectural styles have similarly been strongly influenced by the advent of television and internet access. Urban migration has also contributed to the alteration of rural architecture as young people return home with dreams of living in a more modern house. These factors have led to a preference for modern housing design, predominantly composed of brick and concrete, with features such as large glass windows and tiles. These styles of building are becoming ubiquitous even in remote areas. Unfortunately, these buildings can fail to provide either climatic responsiveness or cultural functionality. For example, Abu-Ghazzeh (1997) investigated architecture in the hot desert climate of Saudi Arabia where the average temperature in summer is 45°C and temperatures as high as 54°C have been recorded (Weather Online, 2019). Here, due to the emerging preference for modern glass and steel architecture, the indoor temperature of buildings is even hotter than the extreme external temperatures, making it essential to use air conditioning systems. The region is also predominantly Islamic, a culture that requires high levels of privacy, and yet the modern housing style has many large windows exposing housing interiors to the outside world.

The loss of vernacular housing design is often lamented in architectural discourse; however, these writings are the feelings of academics, most likely living in urban environments, and do not always represent the feelings of residents in rural areas. During our research we discussed the benefits of vernacular architecture with people from a rural community of Assam, India. Most of the people we spoke to did not feel sentimental attachment to the vernacular design, as to them it symbolised low socio-economic status. Also, while vernacular architecture does have some benefits such as climatic responsiveness, there are advantages to modern housing design features.

There is a lot of research comparing vernacular and modern housing design, however, these are not necessarily two mutually exclusive things. History is not a snapshot, and of course vernacular architectural styles have been constantly evolving over time. In fact, in some regions the alteration of buildings is a part of the cultural tradition. For example, Zetter and Watson (2006) report that impermanence is one of the core ideals of Buddhist philosophy. In Bhutan, a strong Buddhist culture, this belief resonates in their vernacular architecture, where buildings are periodically renovated. There is no obligation to retain any of the features of these houses during the renovation. Residents are free to alter the house to suit their own needs. This process of constant evolution means traditional vernacular buildings can be combined with modern design features.

One of the main drivers behind the growing desire among architects to recapture vernacular design is its adherence to environmental sustainability. Buildings are reported to account for approximately 40 per cent of global energy consumption, energy that is still largely produced by burning fossil fuels. Awareness of the negative environmental impact of burning fossil fuels, accompanied with a growing realisation that their supply is finite, is driving change in our approach to energy use. Thus, smarter building design has the potential to significantly reduce energy consumption and its associated negative environmental impact. The loss of vernacular architecture styles and advent of modern housing in rural areas is increasing building energy consumption rather than reducing it and is therefore counterproductive. The first section of this chapter explores some of the most commonly found climatically responsive housing features. It will end with a discussion on what combination of modern and vernacular features can be combined to create SMART houses.

Many rural residents in developing economies are self-employed or work from home. Thus, their ability to generate income will be impacted by the layout of their house. Ensuring that houses have adequate space for work to be conducted, such as areas that are bright and thermally comfortable, will maximise productivity. Many rural communities also require a specific housing layout to allow them to practise cultural traditions. Figure 3.1 shows the integration of nature, culture and critical regionalism in rural housing design. While ensuring the functional requirements of people are incorporated into their housing design is important, integration of nature and culture in relation to regional or rural characteristics is also an important

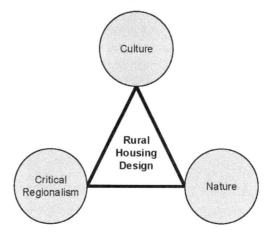

Figure 3.1 Nature–culture–critical regionalism of rural housing design

consideration. The second section of this chapter focuses on housing features that allow functionality of living.

Figure 3.2 shows some of the key elements associated with the functional requirement of rural houses. The elements are income generation, space requirements for primary activities, practice of cultural tradition, appearance and cultural reflection, climatic responsiveness and local skills and resources. Incorporating all of these considerations into the design of a house will aid in making it affordable by going beyond minimising the costs of the house itself to encompass the broader cost of living.

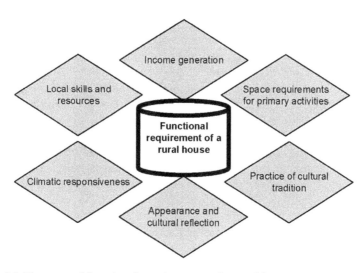

Figure 3.2 Elements of functional requirements of a rural house

Research into vernacular architecture is often conducted from a position of observation, rather than from interaction with the designers and builders. In some regions, this is the only way they can be studied as local knowledge of the traditional building techniques has been lost over time. These studies tend to focus on either the climatic responsiveness or the cultural functionality, with an assumption that the features were designed in response to one or the other of these things. However, interaction with local designers and builders can show a very different picture. For example, in Malaysia many vernacular houses have roofs that curve upwards at the ends (Cromley, 2008). To an architect this appears to be a climatic responsive feature that allows dissipation of heat maintaining cooler internal temperatures in the hot and humid climate. However, when a local builder was asked about this he laughed and said the shape of the roof was supposed to represent the horns of the water buffalo. Thus, vernacular features can simultaneously be designed in response to both culture and climate. Therefore, the chapter will conclude by discussing the overlap between climatic response and cultural functionality.

Climatic responsiveness

Buildings are estimated to account for 40 per cent of energy consumption and the associated greenhouse gas emissions, worldwide. There are significant emissions in both the construction and operational phase of buildings. We will discuss issues relating to construction in more detail in Chapter 6. In this section we consider features that improve the environmental performance of buildings during their operational lifetime. During the operational life of a building achieving thermal comfort is the major consumer of energy followed by lighting and other electrical appliances. The operational greenhouse gas emissions of buildings can be calculated by determining the floor space, amount of energy required per unit floor space, and type of fuel used to generate the energy as shown in Equation 1.

Equation 1

$$E_b = NFS \times EE_{BS,t} \times CI$$

E_b = *CO$_2$ emissions*
NFS = *net floor space*
$EE_{BS,t}$ = *energy efficiency*
CI = *carbon intensity*

(Bauermann and Weber, 2011)

While using low emission energy sources is one option for reducing the negative environmental impact of buildings, improving the energy efficiency will be even more practical as the total number of buildings will continue to increase exponentially and thus it becomes not just an issue of using better fuels but ensuring adequate fuel supplies are available to generate the amount of energy required. As the Earth's limited supplies of fossil fuels is slowly becoming depleted, the cost of these fuels is increasing. In remote areas, the situation is exacerbated by the costs of transporting fuels long distances. Similarly, the efficiency of transporting electricity from a central power station to remote rural areas is very poor as electricity is lost during transmission along power lines (Wirfs-Brock, 2015). The total electricity lost between a power station and its customers depends on the distance (Mahapatra and Dasappa, 2012), hence the further the distance the more energy is wasted. Thus, reducing the operational energy consumption of buildings will not only improve their environmental performance but will also be more economically sustainable and thus increase their affordability.

In warmer climates, the loss of vernacular buildings has led to increased reliance on ceiling fans and air conditioners, meaning increased electricity consumption on hot days. However, in many developing regions, electricity supplies are limited, thus the added drain of many air conditioners often results in blackouts. The houses that rely on air conditioning and ceiling fans for cooling effect are often left with no thermal controls. Exposure to extreme heat can lead to many health problems including dehydration, heat stress and heat stroke. It can also exacerbate underlying poor health conditions especially heart, lung and kidney diseases. For example, a heatwave in Victoria, Australia resulted in a 2.8-fold increase in cardiac arrests (Doctors for the Environment Australia, 2016). There is also some indication that vulnerable members of the community such as the elderly and preterm or underweight infants are also at increased risk. This is very concerning in impoverished developing regions where many children are already suffering from chronic malnutrition. There is also some epidemiological evidence that exposure to extreme heat negatively impacts mental health.

There are other indirect consequences associated with extreme heat. The increased number of health-related incidents will lead to increased stress on the often limited resources of health services, particularly emergency response. The blackouts will also mean loss of refrigeration which will cause some food and medicines to go off. As a result, these will either need to be thrown out, or if consumed could lead to other health problems such as food poisoning. Moreover, blackouts can cause interruptions to transport services.

In regions that experience colder temperatures, vernacular houses are designed to heat up quickly and retain heat. These households are often less dependent on external energy sources, as many still practise open biomass burning. The limiting factor in this case is the availability of fuels. Residents will often need to collect biomass manually, which can involve many hours of walking whilst carrying a heavy load. Thus, they are naturally encouraged to use the fuels efficiently. However, modern building designs lack the insulating properties of vernacular buildings leaving residents more exposed to cold weather.

Almost everyone who has lived in a cooler climate has had first-hand experience of the negative health impacts in the form of the common cold.

Some strains of this common virus can lead to the more severe influenza or even pneumonia. The precise reasons behind the increased occurrence of viral infection during cold weather are still being discovered. Although it was once thought that a direct link existed between being cold and experiencing cold and flu symptoms research has found the reason to be more complex. Firstly, during the winter we are exposed to less sunlight, due to the combination of spending more time indoors, fewer hours of daylight, and increased cloud cover during the day. These factors result in reduced levels of vitamin D, which helps maintain a healthy immune system (Eske, 2018). It has also been proposed that during cold weather the heart compensates by pumping the blood around your body faster to try and keep you warm. This reduces the amount of blood received by your immune system further contributing to its weakness. In extreme cold, the increased strain on the heart can lead to heart failure resulting in hospitalisation or sometimes fatality (Radcliffe, 2018). Other health impacts of extreme cold include frostbite and hypothermia. Therefore, the ability of houses to maintain a comfortable temperature in cold climates is important for maintaining the health of the inhabitants.

Climatic responsive features of vernacular architecture is a popular research field (Zhai and Previtali, 2010). Literature documenting the loss of vernacular architectural styles goes back as far as the 1960s when Rudofsky (1965) noted that modern architecture idolises modern design and aesthetics while ignoring the vernacular, described as homes built by ordinary people for their own needs. Vernacular homes contain a living history of cultures as well as innovative solutions that allow building in irregular shaped locations and housing that naturally provides thermal comfort. Kumar's (1962) thesis noted that vernacular architectural features found in the Punjab region of India could be combined with modern features to provide a model for optimised house design. More recently, vernacular architecture research has become focused on features that could improve the environmental perfor-mance of modern buildings. For example, Bodach, Lang, and Hamhaber (2014) looked at vernacular housing in different climatic regions of Nepal, while Dodo et al. (2014) looked at houses in the Sukur kingdom of Nigeria.

Some of the research into the thermal comfort of vernacular buildings involved surveying occupants. It has been argued that, rather than a result of clever building design, the apparent thermal comfort reported by these people, who are often of low socio-economic status, is simply a result of higher toler-ance for extreme weather conditions. The fact that they are often engaged in agricultural activities as their main source of income means they spend most of their day outdoors and would not experience their home's interior during the hottest part of the day.

Figure 3.3 depicts weather conditions and degree of tolerance by two broad community groups, high socio-economic community and low socio-economic community. For example, Gautam (2008) surveyed residents of Jarkhand, India on their thermal comfort levels and found that they felt comfortable at temperatures outside the thermal comfort temperature ranges

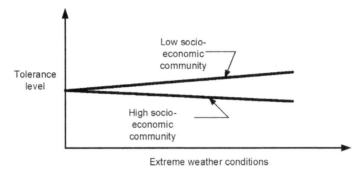

Figure 3.3 Extreme weather conditions and degree of tolerance

defined by the American Society of Heating, Refrigeration, and Air-conditioning Engineers (ASHRAE), the industry standard. Shastry, Mani, and Tenorio (2014) similarly found self-reported levels of comfort of residents of south India falling outside this range.

Dili, Naseer, and Varghese (2010a, 2010b) surveyed people living in both traditional and modern houses in Kerala, India and found people in the modern houses were rarely comfortable inside during the hot summer months, while those in the traditional houses were fine. To ensure that this was not simply the result of vernacular housing dwellers having a higher tolerance for the extreme heat, and the fact that they spend more time outdoors during the day, they also took empirical measurements of indoor temperature, humidity and air movement. These results during the hot summer months found that while the outdoor temperature varied from 22° C to 34°C, the indoor temperature only varied from 26°C to 30°C. Thus, the indoor temperature remained comfortable despite the high fluctuation in outdoor temperature. They also found constant airflow of at least 0.5 m/s helping to minimise the feeling of humidity.

Figure 3.4 shows the key characteristics of vernacular architecture in hot and humid climatic conditions. Climatic responsive features are of course location dependent. Broadly, housing in warmer climates retains coolness during the daytime and cools down quickly in the evening, while in cooler climates housing is designed to retain heat. Some of the specific climatic responsive architectural features commonly found in vernacular structures include:

Orientation: In hot/humid climates buildings are oriented to minimise exposure to sunlight, while maximising exposure to the predominant wind direction (Nguyen et al., 2011; Indraganti, 2010). Windows are positioned to avoid direct sunlight during the hottest part of the day, preventing the internal space from heating up, while facing any ventilation points to the predominant wind direction maximises the potential for natural ventilation. In colder climates, the opposite orientation is employed. Windows are oriented to maximise exposure to sunlight and minimise exposure to winds (Supic, 1982).

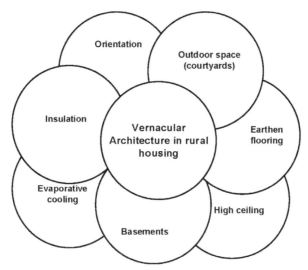

Figure 3.4 Key features of vernacular architecture in hot and humid climate

Insulation: In both hot and cold climates the use of insulation in the roof and walls can help maintain indoor comfort. In warmer climates the thickness of the insulation needs to be managed so that it prevents the house from heating up in the first place, but also so that it allows the heat to dissipate rapidly once the sun has gone down in the evening. In colder climates, the insulation is there to maintain any heat generated indoors thus, it should minimise heat dissipation. As fuels for heating are often limited, the building design allows heat to be retained for as long as possible.

Ceiling height: In warmer climates houses tend to have higher ceilings. As hot air rises, having plenty of space at the ceiling level will allow the heat to gather here. Meanwhile, ventilation points closer to the living area will allow cooler air to enter. In colder climates, the reverse exists. Ceilings are kept low to keep warm air close to the living space. Keeping the total volume of the internal space as small as possible is preferable for maintaining indoor warmth (Ruda, 1998). Ruda (1998) notes that in Hungary, the preference for modern houses includes a "bigger is better" attitude. Yet, much of this indoor space is underutilised. The increased internal space makes the house more difficult and expensive to heat, while taking up a greater chunk of land. Thus, there is less exterior space and less room to grow flora per house.

Basements: These are used in both warm and cold climates. In warm climates they provide a protected cool internal space. Foruzanmehr and Vellinga (2011) describe how spending time in a basement once formed part of the daily routine of households in the warm climate of rural Iran. They describe this concept as "internal nomadism" where people shift themselves around the house to perform activities depending on the time of day or the season. Unfortunately, this practice is extremely unpopular in modern lifestyles for two reasons. People prefer to perform activities in the same space each day. This is possibly due to our modern reliance on appliances to assist with our daily activities, which cannot easily be moved from one room to another. It is also unappealing as basements are often damp and of course are very dark. While the idea is that basements maintain coolness through lack of sunlight, people prefer to spend the daytime hours in naturally lit areas. Basements are also associated with vermin, which are attracted to their damp and darkness. Foruzanmehr and Vellinga (2011) found the use of basements and internal nomadism one of the main reasons cited for abandoning vernacular architectural styles in rural areas of Iran.

In colder climates, building houses partially underground is also a common practice. In Iceland this is reportedly due to the very strong winds that are experienced in the rural areas (Supic, 1982). In regions that were formerly part of the Ottoman Empire internal nomadism was also practised, although unlike in hot/humid climates, this was seasonal rather than daily due to the extreme difference in temperatures experienced in different seasons (Supic, 1982).

Shading: This is also used in both warm and cold climates. In warm climates shading is used to block heat from the sun reaching the indoor space, while in colder climates shading can divert snow and rain from the house to prevent moisture reaching the internal space. Build-up of snow on roofs can be heavy and so having a sloping roof that overhangs the walls is essential to prevent collapse and divert moisture. In temperate climates, such as Japan, where summers are very hot, and winters are very cold, shading can cover both seasons (Kimura, 1994). Due to the higher position of the sun in the sky during summer, carefully positioned window shades will block heat during the summer, but maximise sunlight and warmth during the winter, while also diverting snow and rain. Figure 3.5 shows a vernacular structure from Assam, India, highlighting the use of shading. As seen, natural vegetation is effectively utilised as shading especially against the prevailing sun in summer season.

In some regions, the use of shading goes beyond maintaining thermal comfort for individual households to provide village level benefits. Indraganti (2010) notes that houses are sometimes arranged so that they shelter each other as well as the pathways between buildings where people walk. The large overhangs that many buildings employ to shade their windows, also provide a shaded outdoor space where activities can be performed. Shading is also commonly employed in communal village spaces such as markets.

Landscape features are also used to help maintain thermal comfort. Trees can be strategically planted to provide shading in warmer climates, while in stormy areas they can provide protection from strong winds. Measurement of wind speeds showed that they were reduced by up to one-third by the use of a line of trees (Kimura, 1994). Deciduous trees are good for temperate regions, as during the warmer months, when they are in full bloom, they will maximise shading of windows, while in the winter after they have lost their leaves, they will allow more sunlight to enter the windows while still providing a windbreak during storms.

Figure 3.5 Tea garden bungalow in India

The use of trees to protect houses from the sun and reduce cooling require-ments has been embraced in modern urban development. For example, in Sacramento, California, the local authority initiated a tree planting scheme to help reduce air conditioner use in the early 1990s, which was carefully monitored for over 20 years. The scheme definitely led to a reduction of air conditioner use, although the reduction was less significant than projected. The main problem was, while the council provided the seeds and saplings, the local residents were then responsible for maintaining the health of the trees. Unfortunately, some people failed to tend the trees and they in turn failed to reach their growth potential and, in some cases, died. Dead trees were not replaced with new ones and so they were lost to the scheme (Ko et al., 2016).

Windows: Natural lighting can enhance internal spaces making them safer and more practical for regions with limited electricity access. However, in warmer climates windows can allow heat build-up and in rain or snow can increase the risk of moisture build up. Therefore, windows should be accompanied by some form of shading in all climates. Placing windows high up on a building prevents excess heat building up during the summer, and in the winter, they are essential as snow often piles up quite high making low placed windows redundant (Kimura, 1994).

Courtyards: These are a common feature of vernacular architecture in warmer climates. Courtyards provide an enclosed outdoor space with shelter from the outdoor conditions. Dili, Naseer, and Varghese (2010a) measured temperatures in and around a vernacular house in south India and found that in the lowest part of the courtyard the temperature was 5°C cooler than the outdoor temperature, while in the higher part it was 1.5°C cooler. Thus, they can provide a thermally comfortable private space.

Evaporation: Another natural way to maintain cool interiors is the use of evaporation. The presence of an internal water body, either inside the house or in a courtyard, can provide a source of cool air as the evaporated water is picked up by the breeze (Foruzanmehr and Vellinga, 2011). This passive cooling technique is only effective in arid climates as high humidity will limit the cooling effect of water. In Japan, buildings are elevated, leaving a crawl space at ground level. The purpose is to reduce the risk of flooding but due to the high moisture content of the ground, during the summer evaporation in this space provides a source of cooling (Kimura, 1994).

Despite over 30 years of research into the benefits of bioclimatic features of vernacular architecture, these are still not being widely employed. In some areas vernacular architectural styles have been retained through regulations. Unfortunately, the motivation behind this regulation was often a desire to retain the aesthetics of a given region for the purposes of tourist attraction. For example, Anna-Maria (2009) reports that new housing in Sernikaki, Greece was required to "look" like the traditional vernacular housing but was not required to include the architectural features that were developed for climatic responsiveness. Thus, although these buildings have a vernacular appearance, for all intents and purposes they are modern houses.

The reason for the rejection of vernacular design is often cited as being the desire to emulate urban lifestyles; however, there is little research available to confirm this hypothesis. We have already noted low socio-economic status and internal nomadism as two reasons for rejection of vernacular housing (Foruzanmehr and Vellinga, 2011). At the University of Melbourne researchers visited Majuli Island, a remote, rural region of Assam, India and spoke with residents from the local community about their housing preferences. Most of them preferred modern house designs to the vernacular ones. The reasons were mainly the perception that modern houses looked safer, more resilient and nicer to live in. For example, in this region the vernacular houses often had no windows in a strategy to keep out the heat, but this meant the interiors were very dark. Another resident who had lived in both modern and vernacular houses said that in the vernacular house, the weakness of the walls meant he was unable to hang pictures. Figure 3.6 shows a modern house in Mishing community built on a stilt foundation.

So far, we have focused on the positive features of vernacular architecture; however, there are advantages to modern buildings. Vernacular buildings often lack sanitation facilities and this is a potential contributor to the fact that diarrhoeal disease persists as one of the leading causes of death in developing economies (World Health Organisation, 2018). Incorporating bathrooms into vernacular buildings can be challenging as the building materials can be more susceptible to moisture, quickly developing rot.

Figure 3.6 A typical Mishing house on stilt foundation in a flood prone area in Assam

Agricultural households will often need to store grain indoors and increased interior moisture could also cause this to rot costing the family part of their food supply and potentially their annual income. The natural building materials used in vernacular houses are also more flammable than brick and concrete, making them risky, especially due to the widespread use of open biomass burning. In warmer climates, vernacular buildings minimise the use of windows to keep out the heat, but this makes them dark inside. While this may have been fine for traditional agricultural families, who spent most of their days outdoors, modern lifestyles perform more activities indoors during the day. The lack of natural light will increase the requirement for artificial lighting, and the associated problems with lack of electricity access. Many regions use kerosene lamps as an alternative to electric lights, however, there are many problems associated with these: they are not as bright (United Nations, 2019b), their use has been linked to lung disease, they have the potential to cause fires and explosions, and kerosene is also very expensive (Lam et al., 2012). Therefore, looking at ways of incorporating more natural lighting could help improve quality of life.

The longevity of vernacular buildings has also been raised as a concern. We will discuss this in more detail in later chapters, suffice to say that if houses frequently need to be rebuilt, they cannot be considered sustainable.

This is especially the case for regions which experience frequent extreme weather events, such as Majuli Island, Assam, India, where flooding is an annual event. Therefore, ensuring buildings are designed for resilience is essential and will be described in more detail in Chapter 9.

Another hindrance to revitalising vernacular housing is that because the building techniques were passed down from generation to generation by word of mouth, and never recorded anywhere, the knowledge and skills needed to construct buildings in this style has been lost. In Egypt, Fathy (1973) wanted to bring back the vernacular architecture of domed roof mudbrick houses that had once been the norm. Initial attempts to recreate these buildings through examining the remaining examples of vernacular buildings were a failure. Fortunately, they managed to find some elderly residents in a remote village who had been taught the techniques and still remembered them, thus they were able to document these. However, it may be too late in some regions.

Thus, in many rural areas the traditional vernacular building styles are being replaced with modern brick and concrete houses. Architects see this as a tragedy, due to the potential for their climatic responsiveness features to reduce building operational energy consumption. Residents of rural communities on the other hand, see this as an improvement of their developmental status. Both sides have valid arguments as neither house style is perfect. In modern houses it is more difficult to maintain thermal comfort, while vernacular houses can lack modern facilities such as sanitation that can improve development. A better approach would be to design an optimised house that takes the best features of both vernacular and modern. This idea has been considered previously (Dodo et al., 2014; Kumar, 1962). This ultimate house need not be confined to rural areas but could also benefit urban places.

Cultural design features

While the climatic responsiveness features of vernacular architecture can aid affordability by creating thermally comfortable spaces at minimal cost, it is only one piece of the picture. The internal layout of the house also needs to be functional for the lifestyle of the inhabitants. Houses are multipurpose buildings. At a basic level they provide a place to sleep, but for modern rural families they may also need to function as a workspace, a place to study, a place to entertain guests, a place to practise religious rituals and many other activities. Where the layout is not functional, it may hinder people from improving their development status, or in other cases householders may attempt to make alterations. This can damage the house, reduce resilience, or could even make the house unsafe to live in. In this section we consider three functional requirements of rural households: cultural practises, income generating activities and educational activities.

Religion can play a significant role in the daily activities of people in some cultures. For example, in Islamic culture, entertaining guests in your home is very important. However, privacy is also important. When there is a visitor at the door, they should not be able to see the interior of the house at all from the doorway. If they are invited in, there should be a room for entertaining guests from which the other rooms in the house cannot be seen. Despite this, an affordable housing scheme in Egypt, a predominantly Islamic country, built houses with front doors that opened into the main living area, which comprised of a single internal space (Gelil, 2011). There are various other examples in the literature of housing projects in Islamic countries failing to provide appropriate levels of privacy (Cromley, 2008; Gelil, 2011; Ghaffar-ianHoseini et al., 2011; Abu-Ghazzeh, 1997).

A similar type of segregation is found in the vernacular houses of the Dimasa and Labang tribes of Assam, India (Personal Communication, Velyne Katharpi). These houses typically have an outer room for entertaining guests, and for guests to sleep in should they be staying overnight and an inner room for family members only. Gendered segregation, that is, separate areas for male and female is also a common feature of many vernacular houses including the Dimasa, Labang and Karbi tribes of Assam (Personal Communication, Velyne Katharpi) and Islamic (Gelil, 2011). Figure 3.7 shows a traditional Karbi house in Karbi Anglong district in Assam. In particular, male and female adolescents are often required to have separate bedrooms. Even

Figure 3.7 A typical Karbi house in rural Assam

in some secular developed countries, as we will see in Chapter 4, separate bedrooms for male and female adolescents forms part of the affordable housing policy requirements. Unfortunately, affordable housing schemes often provide a building with a single internal space. While this may suit some cultures, such as the Kuki tribes of Assam, for cultures that require gendered segregation it could either impact on the cultural practices or else the residents may attempt to alter the layout. In trying to alter the building residents have been observed to remove walls and reconstruct with scrap materials which reduce the insulation capacity and resilience of their building and could make it dangerous to live in (Gelil, 2011).

Cleaning habits can also be strongly rooted in cultural influences in ways that affect housing design. For example, in the U.K., most houses have laundry facilities in the kitchen, whereas in Australia, many people reel at the idea of cleaning clothes in the same space where food is prepared (Lawrence, 1982). Here, it is more common for houses to have a separate laundry room, dedicated to cleaning clothes. In many developing rural regions, it is unusual for houses to have a laundry facility at all. Clothes are taken to a nearby waterway, such as a river, and cleaned by hand. While providing these households with a cleaning facility may seem like a developmental improvement, the women, who are predominantly tasked with clothes cleaning, often do not appreciate this. For them, visiting the river to clean the clothes was a rare opportunity to get out of the house and socialise with friends. This highlights the importance of understanding what people want from their house, rather than assuming to know what they need.

The location of the bathroom is perhaps the most predominant culturally biased cleaning issue. In Islamic culture the combination of maintaining privacy while entertaining guests means the bathroom should be located in middle of the house to allow guests to use it without having to pass through private areas of the home such as the kitchen (Gelil, 2011). In other cultures, locating the bathroom away from the main living areas of the house is preferred to prevent moisture build-up and the perception that it is more hygienic. Vastu Shastra, the Hindu system, which integrates architecture with nature, has traditionally believed that locating a bathroom to the northeast of a living space is bad luck. Many architects in India consider this an outdated superstition and will not factor this in when determining the location of the bathroom. Where believers in the Vastu Shastra system end up living in the house, the residents have attempted to remove it (Dili, Naseer, and Varghese, 2010a). This has caused damage to the property while also removing this important sanitation facility.

In some regions, vernacular houses have not traditionally contained a bathroom. Instead washing activities take place in another space. The courtyards typical of vernacular structures in warmer climates often serve as a general purpose space where bathing and cleaning clothes take place among many other activities (Indraganti, 2010). When modern houses replace vernacular ones, the inclusion of an indoor bathroom is seen as a developmental improvement. In India, for example, the various forms of the housing for all initiative, which will be described in more detail in Chapter 7, incorporate bathrooms into the new house design to aid attainment of Sustainable Development Goals. However, this will only be the case if the residents are comfortable using it. Difficulties have also been faced in getting people to use indoor toilets when they are used to practising open defecation (Doloi, Green, and Donovan, 2019). Gastrointestinal disease is still one of the leading causes of death in developing regions, which has been attributed to coming into contact with faecal matter (World Health Organisation, 2018). Unfortunately, there is a lack of awareness of this fact in these regions. Ensuring that toilets are used appropriately is essential to generate real improvements in health and quality of life.

In many rural communities extended families often live in close proximity to offer a support network for one another. In some regions, this has influenced the design of vernacular houses to be comprised of several separate living spaces that are connected by an internal courtyard (Fereig and Al-Khaiat, 1989; Cromley, 2008). This would give individual families privacy while still allowing support and interaction with extended family members. This can be influenced by religious beliefs, for example, for families that practise purdah, it gives women an outdoor space to interact with sisters, cousins and so on without fear of running into anyone else. Access to land can be a limiting factor to the continuation of this tradition in some regions including Kuwait. Fereig and Al-Khaiat (1989) considered the possibility of building houses that would expand upwards, adding a storey for the new family unit, rather than extending the footprint of the house. These types of innovative solutions could allow the continuation of these important cultural traditions. However, the ability to build multi-storey houses will be dependent on other practical factors such as building materials and their ability to support the added weight of multiple stories.

Thus, there are many ways that cultural traditions influence housing design. These cultural practices vary not just across regions but even within a village; different cultural requirements can occur at the household level. In fact, defining the concept of culture can in itself be a challenge without even beginning to attempt to translate that culture into an architectural design feature. Designing an affordable house typically involves creating a simple design that can be implemented in multiple locations as a way to keep costs to a minimum, increasing the difficulty of incorporating cultural requirements into the design. Incorporating culturally appropriate features in housing design may not always be feasible. For example, in the state of Assam, India the state government is responsible for developing an affordable housing scheme. There are 26,000 villages, with more than 100 different ethnic groups with their own cultural needs.

Various authors have considered approaches to incorporating cultural values into architectural design. For example, Rapoport (1987, 1998, 2000) suggested posing the question: "What effects do what aspects of what environments have on what groups of people under what circumstances and why?" Zetter and Watson (2006, 22) suggested that "a nation should be considered a community of diverse cultures, each remaining autonomous for culture specific issues and united for national issues". Figure 3.8 is a Japanese house built with local materials. Figure 3.9 is an example of the vernacular architecture of a typical Japanese double storey house. As seen in both of these houses, the tradition of the community including distinct culture is intact in the design.

An important factor to consider is that even cultural features of vernacular architecture have been evolving over time. Therefore, even if a housing solution is appropriate for today, it may not be in the future. The best approach therefore appears to be a design that incorporates as much inbuilt flexibility as possible. (Gelil, 2011) reports that in Japan small affordable housing units are fitted with removable sliding partitions, as shown in Figure 3.10. Thus, the partitions can be used to create privacy when entertaining guests and removed or altered later to create bedroom spaces for various family members ensuring the option of segregating spaces for different genders. Innovative solutions like this will be the key to providing culturally appropriate affordable housing.

Income generation

In developing rural communities many residents are self-employed. Although this is largely in agricultural related activities, other small businesses also

Figure 3.8 A typical Japanese house with local materials

Figure 3.9 Vernacular architecture of a typical Japanese double storey house

operate. Self-employed people often work from home, especially in rural areas, where lack of commercial building space combined with very limited income would make it impossible for most people to rent business premises. Our data from Majuli Island showed nearly 50 per cent of adult male residents' primary occupation was either farming their own land or business. These requirements have influenced vernacular housing design in various ways.

Livestock farmers need space to house their cattle to prevent them wandering off, and to protect them from predatory wild animals or theft. Farmers therefore often prefer to keep them in their house, where they can keep an eye on them. For example, the vernacular houses of the Sherpa tribe of Nepal are two storey buildings with the livestock occupying the ground floor, while people reside in the upper level (Bodach, Lang, and Hamhaber, 2014). While this provides a safe space for the livestock, they also provide a thermal buffer zone during the harsh cold winters experienced in this region. In Assam, some of the vernacular houses are built on stilts, to cope with the frequent flooding that occurs in this region. This space also acts as a protected place to house livestock.

Figure 3.10 Sliding door in Japanese house

Arable farmers will need a space to store grains that is cool and dry to prevent mould formation and protected from vermin and other pests. The vernacular houses of the Tamang tribe of Nepal are two storeys, similar to the Sherpa tribe, but in this case the ground floor forms the main living area while the upper floor is used to store grain (Bodach, Lang, and Hamhaber, 2014). Arable grain often needs to be dried after harvesting. In developing rural regions, due to the lack of energy sources this is generally done in the open air, putting the grain at risk from vermin, moisture and so on. Therefore, courtyards often act as a protected space for this practice (Indraganti, 2010).

Education

Obtaining an education is one of the most important methods for breaking the poverty cycle in rural communities. While the United Nations Millennium and Sustainable Development Goals have done much to ensure children have access to a school (United Nations, 2019a), to achieve effective educational attainment, especially in secondary school, they also need an appropriate space to study outside school hours. A secondary school student from the Pacific Island nation of Tonga noted that sometimes other children would remain in the classroom after school to work on the homework assignments (Falemaka, 2019). She asked them why they didn't go home to do that, and they responded that their house did not have any quiet space for them to study. She believed that this led to many children dropping out of school and eventually turning to illegal activities to make money as adults. Thus, a home should provide a quiet, well-lit area for school-age children.

Nurturing cognitive development of younger children, particularly under 5s, has been shown to improve long-term well-being (Lo, Das, and Horton, 2016; Perez-Escamilla and Moran, 2017; Richter et al., 2017). Being engaged in learning stimulation activities during this crucial developmental stage is important for long-term quality of life, thus stimulating play spaces should be provided for very young children. Where children of multiple ages exist, as is the case in many rural villages, having separate spaces for children of different age groups is therefore important.

Architecture as a living history

Other aspects of cultural architectural features have evolved through historical events that occurred in the region. These do not always have a positive backstory, but it has been argued they provide an important living history of the region. For example, in areas that have experienced long-term conflict such as Saudi Arabia, design features illustrate the need for defence (Saleh, 2000). In Iran, the houses of the oppressed Zoroastrian community have had design restrictions placed on their homes (Mazumdar and Mazumdar, 1997). Some of these included restrictions on the height – their roof line had to be able to be reached by an Islamic member of the community. To make the interiors of their homes functional, it was therefore essential for them to dig down and create a living space that was partially underground.

In the dusty desert climate of Iran, having natural ventilation can lead to excess dust entering the house. In response, a cooling system known as Badgirs are in common use. However, Zoroastrian households were not allowed to include this feature. Thus, these households often contain an indoor pond to provide cooling through evaporation.

The use of height restrictions to distinguish the socio-economic status of different households is not unique to Iran. A similar restriction on building height was imposed on lower caste houses in India. In Myanmar, this is also used to distinguish the homes of kings, monks and high officials from other members of the community (Zetter and Watson, 2006). The height of the building was only one of many restrictions placed on ordinary people to distinguish their homes from those of elites. Other things included not being allowed to adorn the exterior with paint, lacquer, gold or sculptures, and not being allowed to use bricks but only lightweight building materials such as bamboo.

Overlap of culture and climatic response

Thus far we have considered bioclimatic and cultural aspects of vernacular architecture as two separate dimensions; however, there is some overlap between these. For example, in the Climatic responsiveness section we described how a courtyard can provide a way to maintain thermal comfort, while in the second section we described how this area is used for bathing, and drying grain. Courtyards also act as places were people entertain friends, local communities gather for meetings, children play and ceremonies such as weddings take place (Indraganti, 2010). In the Zoroastrian community of Iran courtyards were a safe space for people to gather and practise religious rituals without fear of persecution from their Islamic neighbours (Mazumdar and Mazumdar, 1997). Thus, while courtyards were initially developed in response to climatic conditions they have evolved into important cultural gathering places. In colder regions, where there are no courtyards, many villages contain a centrally located building available for the community gatherings and other cultural functions provided by courtyards in warmer climates (Singh, Mahapatra, and Atreya, 2011).

Modern housing designs do not usually contain a courtyard, due to the understanding that such bioclimatic features can be substituted with air-conditioning and ceiling fans. This attitude not only goes against modern sustainability principles, but loss of courtyards also changes local culture by taking away this important communal space. This highlights the importance of understanding local culture when designing housing.

Another example of misinterpretation of housing design arose as architects attempted to design culturally sensitive houses for Majuli Island, Assam. In this region, houses are built on stilts due to the frequent flooding of the region. A log is typically used to access the higher-level living space. Architects from the University of Melbourne suggested replacing this with a staircase, however, a local resident noted that one of the reasons for using a log is that snakes are unable to climb it and therefore cannot access the living space. They would easily be able to slither up the smooth surface of the staircase. Engaging with the local community is essential to ensuring appropriate housing design.

Policy Implications

The optimum design of a house will not involve sentimental retention of vernacular styles or complete modernisation. Instead, it will require a combination of the most important design aspects of both. The best way to ensure appropriate design is to consider the input of end-users with equal weighting to input from architects and engineers. This is the only true way to ensure the layout provides the required functionality. In reality unfortunately, it is likely that only a handful of end-users will be able to participate in the design process, therefore, incorporating flexibility into the design as far as possible will also be essential to ensure all cultures can function in their home. Another school of thought is to provide villagers with the funding and tools to build their own homes, giving them greater control over the final design. An important component of this will be training residents in construction skills. It is hoped that engaging residents in this way will have a trickledown effect, that is, an initial group of villagers trained in construction will be able to train other villagers, leading to self-sufficiency in the community. This philosophy of engaging residents into the development of their home is shared by charitable organisations such as Habitat for Humanity who have learned from many years of working with remote rural communities.

References

Abu-Ghazzeh, T. M. (1997). Vernacular architecture education in the Islamic society of Saudi Arabia: Towards the development of an authentic contemporary built environment. *Habitat International*, 21, 229–253.

Anna-Maria, Vissilia. (2009). Evaluation of a sustainable Greek vernacular settlement and its landscape: Architectural typology and building physics. *Building and Environment*, 44, 1095–1106.

Bauermann, Klaas, and Weber, Christopher. (2011). Heating systems when little heating is needed. In Fereidoon P. Sioshansi (ed.), *Energy, Sustainability and the Environment*. Oxford: Butterworth Heinemann.

Bodach, Susanne, Lang, Werner, and Hamhaber, Johannes. (2014). Climate responsive building design strategies of vernacular architecture in Nepal. *Energy and Buildings*, 81, 227–242.

Cromley, Elizabeth. (2008). Cultural embeddedness in vernacular architecture. *Building Research and Information*, 36, 301–304.

Dili, A. S., Naseer, M. A., and Varghese, T. Zacharia. (2010a). Passive environment control system of Kerala vernacular residential architecture for a comfortable indoor environment: A qualitative and quantitative analysis. *Energy and Buildings*, 42, 917–927.

Dili, A. S., Naseer, M. A., and Varghese, T. Zacharia. (2010b). Thermal comfort study of Kerala traditional residential buildings based on questionnaire survey among occupants of traditional and modern buildings. *Energy and Buildings*, 42, 2139–2150.

Doctors for the Environment Australia. (2016). Heatwaves and health in Australia: Fact sheet. Doctors for the Environment Australia. Retrieved from www.dea.org.au/climate-change-and-health-in-australia-fact-sheets/.

Dodo, Yakubu Aminu, Ahmad, Mohd Hamdan, Dodo, Mansir, Bashir, Faizah Mohammed, and Shika, Suleiman Aliyu. (2014). Lessons from Sukur vernacular architecture: A building material perspective. *Advanced Materials Research*, 935, 207–210.

Doloi, Hemanta, Green, Ray, and Donovan, Sally. (2019). *Planning, housing and infrastructure for Smart Villages*. Abingdon, U.K.: Routledge.

Eske, Jamie. (2018). What's the link between cold weather and the common cold? *Medical News Today*. Retrieved from www.medicalnewstoday.com/articles/323431.php.

Falemaka, A. M. (2019). Youth panel with country representatives from the youth essay competition. Pacific Peoples Housing Forum, 17 May, 2019, Auckland, New Zealand.

Fathy, Hasan. (1973). *Architecture for the poor*. Chicago: University of Chicago Press.

Fereig, Sami M., and Al-Khaiat, Husain. (1989). Building expandable housing units in Kuwait. In Oktay Ural and L. David Shen (eds), *Affordable housing: A challenge for civil engineers*. New York: American Society of Civil Engineers.

Foruzanmehr, Ahmadreza, and Vellinga, Marcel. (2011). Vernacular architecture: Questions of comfort and practicability. *Building Research and Information*, 39, 274–285.

Gautam, Avinash. (2008). *Climate responsive vernacular architecture: Jharkhand, India*. Manhattan, KS: Kansas State University.

Gelil, Nermine Abdel. (2011). Less space, more spatiality for low income housing units in Egypt: Ideas from Japan. *International Journal of Architectural Research*, 5, 24–48.

GhaffarianHoseini, AmirHosein, Dahlan, Nur Dalilah, Ibrahim, Rahinah, Baharuddin, Mohd Nasir, and GhaffarianHosein, Ali. (2011). The concept of local-smart-housing: Towards socio-cultural sustainability of vernacular settlements. *International Journal of Architectural Research*, 5, 91–105.

Indraganti, Madhavi. (2010). Understanding the climate sensitive architecture of Marikal, a village in Telangana region in Andhra Pradesh, India. *Building and Environment*, 45, 2709–2722.

Kimura, Ken-ichi. (1994). Vernacular technologies applied to modern architecture. *Renewable Energy*, 5, 900–907.

Ko, Yekang, Roman, Lara A., McPherson, E. Gregory, and Lee, Junhak. (2016). Does tree planting pay us back? Lessons from Sacramento, California. *Arborist News*, 50–54.

Kumar, J. M. (1962). *Rural housing for India*, Master of Architecture Thesis, University of Manitoba, Canada.

Lam, Nicholas L., Smith, Kirk R., Gauthier, Alison, and Bates, Michael N. (2012). Kerosene: A review of household uses and their hazards in low- and middle-income countries. *Journal of Toxicology and Environmental Health B: Critical Reviews*, 15, 396–432.

Lawrence, Roderick J. (1982). Domestic space and society: A cross-cultural study. *Society for Comparative Study of Society and History*, 104–130.

Lo, S., Das, P., and Horton, R. (2016). Early childhood development: The foundation of sustainable development. *Lancet*, 389, 9–11.

Mahapatra, Sadhan, and Dasappa, S. (2012). Rural electrification: Optimising the choice between decentralised renewable energy sources and grid extension. *Energy for Sustainable Development*, 16, 146–154.

Mazumdar, Sanjoy, and Mazumdar, Shampa. (1997). Intergroup social relations and architecture: Vernacular architecture and issues of status, power, and conflict. *Environment and Behavior*, 29, 374–421.

Nguyen, Anh-Tuan, Tran, Quoc-Bao, Tran, Duc-Quang, and Reiter, Sigrid. (2011). An investigation on climate responsive design strategies of vernacular housing in Vietnam. *Building and Environment*, 46, 2088–2106.

Perez-Escamilla, R., and Moran, V. H. (2017). The role of nutrition in integrated early child development in the 21st century: Contribution from the *Maternal and Child Nutrition Journal*. *Maternal and Child Nutrition*, 13.

Radcliffe, Shawn. (2018). How extremely cold weather can affect your health. HealthLine. Retrieved from www.healthline.com/health-news/how-extremely-cold-weather-can-affect-your-health#8.

Rapoport, Amos. (1987). On the cultural responsiveness of architecture. *Journal of Architectural Education*, 41, 10–15.

Rapoport, Amos. (1998). Using "culture" in housing design. *Housing and Society*, 25, 1–20.

Rapoport, Amos. (2000). Theory, culture and housing. *Housing, Theory and Society*, 17, 145–165.

Richter, L. M., Daelmans, B., Lombardi, J., Heymann, J., Boo, F. L., Behrman, J. R., Lu, C., Lucas, J. E., Perez-Escamilla, R., Dua, T., Bhutta, Z. A., Stenberg, K., Gertler, P., and Darmstadt, G. L. (2017). Investing in the foundation of sustainable development: Pathways to scale up for early childhood development. *Lancet*, 389, 103–118.

Ruda, Gy. (1998). Rural buildings and environment. *Landscape and Urban Planning*, 41, 93–97.

Rudofsky, Bernard. (1965). *Architecture without architects*. New York: The Museum of Modern Art.

Saleh, Mohammed Abdullah Eben. (2000). The architectural form and landscape as a harmonic entity in the vernacular settlements of southwestern Saudi Arabia. *Habitat International*, 24, 455–473.

Shastry, Vivek, Mani, Monto, and Tenorio, Rosangela. (2014). Impacts of modern transitions on thermal comfort in vernacular dwellings in warm-humid climate of Sugganahalli (India). *Indoor and Built Environment*, 23, 543–564.

Singh, Manoj Kumar, Mahapatra, Sadhan, and Atreya, S. K. (2011). Solar passive features in vernacular architecture of north east India. *Solar Energy*, 85, 2011–2022.

Supic, Plemenka. (1982). Vernacular architecture: A lesson of the past for the future. *Energy and Buildings*, 5, 43–52.

United Nations. (2019a). Goal 4: Quality education. United Nations. Retrieved from www.un.org/sustainabledevelopment/education/.

United Nations. (2019b). Goal 7: Affordable and clean energy. United Nations. Retrieved from www.un.org/sustainabledevelopment/energy/.

Weather Online. (2019). Saudi Arabia. Weather Online. Retrieved from www.weatheronline.co.uk/reports/climate/Saudi-Arabia.htm.

Wirfs-Brock, Jordan. (2015). Lost in transmission: How much electricity disappears between a power plant and your plug? Inside Energy. Retrieved from http://insideenergy.org/2015/11/06/lost-in-transmission-how-much-electricity-disappears-between-a-power-plant-and-your-plug/.

World Health Organisation. (2018). The top 10 causes of death. World Health Organisation. Retrieved from www.who.int/news-room/fact-sheets/detail/the-top-10-causes-of-death.

Zetter, Roger, and Watson, Georgia Butina. (2006). *Designing sustainable cities in the developing world*. Aldershot, U.K.: Ashgate Publishing House.

Zhai, Zhiqiang (John), and Previtali, Jonathan M. (2010). Ancient vernacular architecture: Characteristics categorization and energy performance evaluation. *Energy and Buildings*, 42, 357–365.

4 Affordable houses

The phrase "affordable house" can mean different things to different people, depending on their socio-economic status and where they live. In many developed nations, such as the U.S., pockets of affordable housing were developed in areas segregated from the rest of the population (Mueller and Tighe, 2007). This has resulted in concentrated areas of poverty where substance abuse and crime thrive. Children born into these areas attend substandard public schools and end up trapped in the poverty cycle. These types of affordable housing schemes have resulted in the term having strong negative connotations. People with medium to high level wealth in the U.S. believe that living in close proximity to affordable houses will bring down the value of their own property.

Thus many proposed affordable housing schemes that would see low-cost housing established in a socio-economically diverse area are rejected due to public opposition. The situation is somewhat exacerbated in rural areas where local government is often comprised of local elites who want to protect their own assets and will therefore only support projects they perceive will enhance their properties' value (Dolbeare, 2001; Tighe, 2010). This perception that affordable houses drive down the price of nearby housing has been found to be false. Nguyen (2005) showed that as long as the quality of the houses meets a standard and that this is maintained, there is no detriment to the value of nearby properties. In fact, where affordable houses are built in run-down areas, such as abandoned former industrial sites, they were shown to increase the value of nearby houses. Socio-economic diversity has also been found to have multiple community benefits as we will discuss in Chapter 5.

While there is some precedence for the negative image of affordable housing schemes, in reality, affordable housing is not solely an issue for the lowest socio-economic class, that is, people falling under the definition of poverty, people subsisting on welfare payments, or earning a wage that is below living standard. In Sydney, Australia a person earning a medium level salary would certainly not be considered poor or qualify for welfare benefits or other government subsidies, yet their income is not sufficient to cover the costs of housing in close proximity to where they work. Many people instead spend several hours per day commuting between their work and their home. The reasons for this are related to a physical shortage of housing coupled with incentives for people who are already home owners to purchase investment properties. This has caused the price of housing to increase at a much higher rate than inflation. For example, in New Zealand it was noted that the mean price of a house is now 22 times higher than it was a few decades ago, while the mean salary has only increased four times over the same time period (Salesa, 2019). The properties purchased as investments could potentially create a thriving rental market; however, these are often left vacant and only used for vacations. Thus, large numbers of houses are sitting empty in locations convenient to employment opportunities and services, while people working in these locations are struggling to find affordable accommodation (Paris, 2007).

People struggling to find affordable housing often includes those working in essential services jobs. These can incorporate long hours or shift work and thus, would benefit greatly from having workers living close by. One approach to help improve the perception of affordable housing is simply to call it by a different name. For one affordable housing project, the scheme was introduced under the title "keyworker" housing (i.e. housing aimed at low-income essential services employees) (Haylen, 2015). Although the description of this project was identical to a previously proposed affordable housing project, there was a much higher level of acceptance from the local community. In the U.S. reframing the title of affordable housing as "workforce housing" or "lifecycle housing" was shown to incur less opposition from nearby residents, despite being an identical policy in every other way (Tighe, 2010). However, the author noted that over time as awareness of the new terminology grows, people will begin to assign it the same connotations as affordable housing, so the terminology has a shelf life and must be constantly updated to reduce public opposition. Thus, simply renaming something can improve the chances of obtaining planning permission.

Thus far we have considered the public perception of affordable housing. From the policy side, defining affordable housing has been debated in academic circles for decades. For this book we work under the definition of affordable housing coined by Maclennan and Williams (1990, 9): "affordability is concerned with securing some given standard of housing (or different standards) at a price or rent which does not impose, in the eyes of some third party (usually government), an unreasonable burden on household incomes." This defines the structure of the rest of this chapter. In the first section we explore different ways that have been developed to define a price or rent that does not impose unreasonable burden. Within this we consider issues pertaining to both homeowners, in terms of the mortgage repayments, and renters.

In the second section of this chapter we look at the "given standard" part of the definition to focus on ensuring adequacy is incorporated into the formulation of affordable housing policy. In the third section we discuss issues relating to the availability of housing that fits in with the affordable housing definition. The final section of this chapter will look at the role governments play in creating affordable housing through various policy instruments.

How to define an affordable house

In many regions, governments offer assistance with housing costs. When determining eligibility for these welfare schemes, an affordable house has been defined by the percentage of the household's monthly income that is spent on monthly accommodation costs. This can be either rent or mortgage payments and usually includes other essential housing costs such as maintenance and thermal comfort (Luffman, 2006). The threshold for eligibility is usually around 30 per cent of monthly income; if any more than this is spent on monthly housing costs the household is eligible for the scheme. While this seems like a logical assumption, the calculation has proven to include people who are actually comfortably well off, while others who are in genuine need miss out.

At the high end, there are many people earning medium to high incomes who choose to spend a significant proportion of their income on their house. There are various reasons for this: in some instances it is related to the location of the property which may offer them more convenient access to services or allow their children to attend a better school. In others they may choose to live in a larger house with better facilities such as a swimming pool. While these people are spending more than 30 per cent of their income on their house, the amount of money left over is still substantial enough for them to afford other basic necessities such as food and clothing and therefore they should not be considered eligible for any affordable housing benefits.

At the other end of the scale people on very low incomes may be spending less than 30 per cent of their income on housing, but the amount of money left over is insubstantial and they may be forced to cut back on essentials. Kutty (2005) devised a new calculation of affordability for the U.S. defined as "housing induced poverty". This new calculation was based on the total amount of a household's income left over, after housing costs had been paid each month. This value is more appropriate because it gives a clearer picture of the amount of money available to purchase other essentials. The average cost of food, clothing and other basic essentials for the number of people living in the house can then be compared to the total amount of monthly income available after accommodation costs have been paid to determine whether they are comfortable or whether they have to go without basic essentials to cover their accommodation expenses.

Figure 4.1 encompasses the key considerations that define the affordability in broad context: *building cost, monthly ownership cost, thermal comfort, income generation potentials, jobs and career, disposable income, accessibility of necessary amenities* and *access to funds*. As the figure depicts, affordability is not defined by mere disposable income but by a raft of other considerations.

Houses are generally either owner-occupied or rented. So far, the affordability calculation has considered both of these together; however, there are some differences that need to be considered when determining a household's

Figure 4.1 Defining affordability

status. Firstly, we consider issues relating to owner-occupied houses. In any world region, whether developed or developing, the ability to purchase a house outright is impossible for the majority of residents. Most people would need to save money for years or even decades to buy a house. This is not an ideal situation because, of course, most people wish to purchase a house when they are younger so they can create a home for their children. Thus, borrowing money is the normal approach to house purchase. Even people on high incomes in developed countries would rarely think of purchasing a house outright as mortgages are readily available for anyone able to prove they have a consistent income. Mortgages are usually considered a lower risk than other types of loans, because the value of a house is more likely to increase over time than decrease. Thus if the borrower fails to make payments, the lender will be able to recoup their money through selling the house.

For most people in developed countries therefore, to buy a house it is only necessary to save enough money to pay a deposit of around 20 per cent of the total cost of the house. While this may sound achievable, many people will be living in rental accommodation while trying to save which can be difficult if rental payments are substantial. Once this deposit is accrued, the rest of the money can be borrowed from a bank or other lending facility and repaid in monthly instalments. These monthly instalments include interest and tend to fluctuate in line with inflation. Thus people on low incomes struggle to gain access to the housing market, either because their rental accommodation costs prevent them from saving money for a deposit or if their mortgage repayments make up a significant portion of their monthly income, the risk of a sudden increase in mortgage interest could be too great.

To cater to these people, mortgage lenders often offer products with greater flexibility, known collectively in the U.S. as "sub-prime" mortgages (Heyford, 2019). In some cases, they will accept a much lower deposit of 10 per cent, such as "dignity mortgages" in the U.S. or even 5 per cent from some banks in Australia. To make up the added risk of taking on these clients the banks charge above market interest rates. In Australia, this added interest goes towards "lenders mortgage insurance" which guarantees payment from a third party insurance company if a client cannot pay their mortgage. The "dignity mortgage" charges extra interest during the first few years of the loan; however, if the client consistently makes payments on time during this period, their interest rates will drop back to the market rate. This product gives clients a strong incentive to stick to their repayment schedule.

Another mortgage product aimed at low income households is "fixed rate mortgages". As the name suggests, these offer a consistent monthly repayment for a fixed period of time, usually a few years. Thus, there is no risk of interest payments suddenly increasing, although it also means that the household will miss out if interest rates suddenly drop. Lenders usually offer these mortgages at a fixed rate that is slightly above current market interest rates to recoup the potential loss of income if mortgage rates do increase during the period of the offer. Thus, clients of fixed price mortgages often end up paying more interest than those on variable mortgage rates. However, it provides a way into the property market for low income households that would not be approved for a traditional mortgage product. Most lenders will allow clients who start out on a fixed rate mortgage to change to a variable rate over time, if they keep up their repayments or increase their income. Other people who may struggle to obtain mortgages are those on variable incomes. This could include self-employed people who may lose money some months to cover business expenses, but make profits later on. For people with these types of inconsistent income, "interest only mortgages" are offered by some lenders. In this case, the monthly repayments only cover the interest and thus are not paying off the loan. However, the product allows for flexible payments, so when their small business is in surplus, they can pay off a large chunk of their mortgage (Heyford, 2019).

Initially, the use of sub-prime mortgage products was very positive as it aided many low-income families into home ownership. Unfortunately, in the early 2000s banks and mortgage lenders saw the potential to make huge short-term profits from these products and relaxed eligibility criteria even further. They also employed predatory lending techniques, including deceptive, coercive, exploitative or unscrupulous actions to convince consumers to purchase a loan that they more than likely would not be able to pay back. Initially, the scheme opened up the housing market to a new sector of the population, thereby increasing demand and housing prices. However, as we saw in 2007–2008, many of these high-risk borrowers suddenly defaulted on their mortgages simultaneously. As a result, the value of housing dropped, and lenders were unable to sell the houses for the price of the loans, leading to the financial crisis (French, Leyshon, and Thrift, 2009). Thus, while the misuse of sub-prime mortgages led to a very negative outcome, if managed responsibly these products can provide a benefit to many low income families.

A more recent approach being tried in the U.S. will take the location of the house into account. Where the house has good access to services and infrastructure, a lower deposit, and higher monthly repayments may be offered on the grounds that the residents of the household will be saving money in other ways. For example, if the house has good public transport links or is located walking distance from the places that members of the household regularly need to attend, it may not be necessary for them to own a car, which will dramatically reduce their monthly expenses freeing up more money for their mortgage repayments (Acolin, 2018). Such more sensible approaches to sub-prime mortgage lending along with better regulation could provide a more genuine way to assist low income households into homeownership.

While access to mortgage products in developed countries is fairly ubiquitous, the picture for developing nations is very different. Hewson (2012) reports that in Africa only the top 5 per cent of the population have sufficient income to secure a mortgage. This is despite the fact that, in many African nations, development is occurring rapidly and there is an emerging middle class that have secure jobs with steady income. These people would be a safe investment for a loan but fall outside the existing mortgage requirements. While the very poor are often eligible to receive government funded subsidies to aid the cost of their housing, it is this emerging middle class that misses out at both ends, with a salary that is too high for government subsidies but too low for mortgage access. The introduction of responsibly managed subprime mortgage products could open up the housing market for many of these people, leading to substantial developmental improvements.

In India, despite various financial inclusion policies since independence from the U.K. many people, particularly in rural areas, do not have access to finance products. A report from 2013 showed that 52 per cent of rural households did not have a bank account, while the 2011 census found that 73 per cent of agricultural households did not have access to credit (Chetia, 2018). An existing bank account and credit history are two of the main required criteria for obtaining a mortgage in India. There were two components to this exclusion. Firstly, many people physically lacked access to a bank as there was none located in their village or the surrounding villages. This coupled with poor transport infrastructure meant many people were simply unable to go to a bank. Secondly, there is a social component to exclusion. Financial institutions tend to favour customers with higher incomes and so often denied lower income households a bank account or credit. Similarly, poor literacy rates are prevalent in rural India meaning many people are unable to fill in the forms required to apply for a bank account or a loan (RGVN, Personal communication, 2017). Many rural communities' only way to access finance products was through assistance from non-governmental organisations such as Rashtriya Gramin Vikas Nidhi (RGVN) who help rural communities set up small business initiatives.

However, things are beginning to change. Since 2014, the Government of India has implemented an initiative to increase rural communities' access to financial products under a programme known as Jan Dhan Aadhaar Mobile (JAM) (Chetia, 2018). Under this scheme a reported 300 million bank accounts have been opened. One of the key features of the scheme is embracing the use of technology, allowing internet banking to make up for the lack of physical banking facilities in many areas. Smart phone ownership is currently around a quarter of India's population; however, it is increasing. Unfortunately, there is a huge gap between male and female access to smart phones (34 and 15 per cent respectively), which has actually widened compared to 2014 when it was 16 per cent and 7 per cent. This is concerning because it will further limit access to banking products for women. Embracing technology to open up finance products to rural communities could benefit other developing regions where smart phone access is high, such as Mexico where it is at 66 per cent and South Africa where it is at 52 per cent (Wiggers, 2019).

The other main way to access accommodation is renting. Renting can be more beneficial for low-income families who are unable to attain mortgage approval, unable to save for a deposit, or do not want to incur the responsibility that goes along with home ownership such as the need to perform regular repairs and maintenance. Owner-occupied homes of lower income people often end up falling into disrepair as the high proportion of their monthly income taken up by mortgage repayments means there is not enough money available for sudden urgent repairs. For example, a broken hot water system usually needs to be completely replaced if it breaks down. The total cost of this replacement which includes the system itself, removal of the broken system and installation of the new system can be in the thousands. While these types of repairs can sometimes be covered by insurance, if they are not many low-income homeowners would not have this amount of money in savings, and thus the system would remain broken leaving the household with no hot water for an indefinite period of time. In rented accommodation the responsibility of these repairs usually falls on the building owner.

The down side of living in rental accommodation is the threat of impermanence. Without regulation, rent payments can be increased by any amount at any time. Thus, a sudden increase in rent could force a family to move. The owner of the building may also decide to sell the property at some point again risking the stability of the family living there. Affordable housing policy targeting people in rental accommodation can therefore be important for allowing stability. These policies can either be targeted to the property owner, for example putting restrictions on rental increases, or the tenants, such as providing financial assistance to cover the cost of the rent.

Figure 4.2 depicts the information flow between the owning and renting of a house from the perspective of affordability. As seen, the affordability of renting or owning a house depends on a range of checks and balances. For

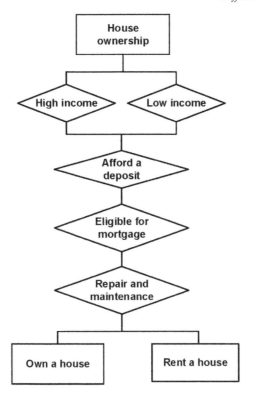

Figure 4.2 Owning versus renting a house

instance, the legal fees associated with transferring home ownership documents can be prohibitively expensive, therefore even after saving a deposit the dream of home ownership is still beyond the reach of some people. Similarly, house ownership comes with the responsibility of maintenance and upkeep of the property and leading a decent life. Unstable income may hinder routine maintenance and essential repairs, deteriorating property values over time. Thus the decision whether to own or rent a house is an important one from an affordability view point.

Placing restrictions on rental increases benefits the house occupants; however, in a free market the property owner may struggle to cover the costs of the property if they incur expenses such as increases in land tax, without being able to pass these costs on to their tenants. Portugal has had long-term rent control legislation in place and provides an example of the dangers of restricting rental increases. In Portugal tenants had the right to indefinite leases which were extremely difficult for a property owner to terminate (Hatton, 2012). Property owners were not allowed to increase the rent until the lease was terminated and a new occupant took over the property. This incentivised tenants to remain in the same property for decades, passing occupancy onto relatives to avoid a rent increase. As a result, property owners were not profiting from their investment and did not have enough money to perform basic maintenance on their buildings. This has resulted in many properties in Portugal falling into a state of disrepair. The severity of this situation has been described as "a textbook case on how to destroy a city without bombing it" (Hatton, 2012).

Two other negative outcomes resulted from the legislation. Firstly, if or when a tenant finally left a property, the owner would increase the rent far above market value, to insure themselves against future long-term residents. Secondly, it has resulted in property owners developing a preference for short term leasing, such as holiday rentals. This ensures a high turnover of tenants, allowing owners to maintain rent at market value, but means fewer rental properties available for local residents. While Germany and the Netherlands have had similar rent control legislation in place, the impact has been less significant. One of the main reasons is that Portugal is a significantly poorer country with minimum wages of $624 per month (compared to U.S. $1317 per month) and average salary of $1030 per month (compared to U.S. $3500 per month).

Portugal is starting to transition out of their long-standing rigid rent control legislation. This is largely in the wake of the financial crisis in the late 2000s, where the country was bailed out on the proviso these outdated tenancy laws be scrapped. Thus, property owners are starting to be able to increase rents. However, the transition is slow as the well-being of renters also needs to be carefully managed. A sudden substantial rent increase could see a family forced to move out of a home they have occupied for decades, and unable to find alternative accommodation in the neighbourhood they call home (Wise, 2017; Hatton, 2012).

The benefit of providing assistance to renters, rather than placing restrictions on property owners has been noted elsewhere. Kutty (2005) compared rental assistance programmes in the U.S. and found that targeting renters rather than property owners led to more successful outcomes. The article also noted that rental assistance programmes should not be a long term solution but should assist people to obtain self-sufficiency. For example, providing people with accommodation close to employment opportunities could allow them to obtain stable employment and become established in a new community. Overtime this will allow them to build up a good credit rating and save for a deposit, hopefully allowing them to enter the property market.

Adequacy

So far in this chapter we have considered housing affordability from a purely monetary perspective, however, as stated in the quote in the introduction, the standard of housing also needs to be given consideration. Morton, Allen, and Li (2004) note that the basic standards should include the issues of: internal physical and health conditions, structural defects, and overcrowding. Similarly, the first target of the United Nations Sustainable Development Goal 11 states "By 2030, ensure access for all to adequate, safe, and affordable housing and basic services" (United Nations, 2019). In this section we look at the concepts of adequacy and safety and how they also need to be incorporated into calculations of affordable housing.

Health and safety

There are many people who would not be considered in need of affordable housing according to their finances, who are living in inadequate conditions. Inadequate houses are not just unpleasant to live in; they can also impact the health and safety of the occupants, both directly and indirectly. Direct impacts include the presence of hazardous materials such as lead paint and asbestos. Particularly when a house is dilapidated the hazardous asbestos fibres can be released into the air leading to inhalation and associated lung disease. As houses fall into disrepair, they lose the ability to keep out moisture, whether from leaking pipes, rain entering through cracks around doors and windows or rising damp due to poor foundations. Excess indoor moisture can lead to mould which can produce allergens that can exacerbate the symptoms of allergy sufferers, including eczema, and asthmatics. Children and the elderly are particularly susceptible to mould-related health problems. Houses in dilapidated states may also have other physical hazards such as splintering wood, broken glass or faulty electrical wiring.

Indirect health impacts include increased susceptibility to diseases. Firstly, stress is associated with a wide range of health impacts. People on low incomes are at increased risk of suffering from stress. This can come from a wide range of sources including struggling to pay bills, living in a poor neighbourhood with high crime rates, poor access to services such as good schools for children to attend. These issues also put people at high risk of experiencing mental health problems (Mueller and Tighe, 2007).

People in low income housing are also at increased risk of malnutrition. This can be a result of the inability to afford adequate food through housing induced poverty. It may also be due to living in a house with poorly maintained cooking facilities, making it difficult to prepare healthy meals; or lack of access to a well-functioning refrigerator making it difficult to store food safely. The combination of inadequately stored and inadequately cooked foods also leads to an increased risk of food poisoning.

As described in Chapter 3, thermal comfort is also essential to good health especially in regions that experience extreme cold or extreme heat conditions. While we have seen that houses can be designed to naturally retain thermal comfort to some level, in reality few houses are being built using these techniques today. Many have inadequate insulation and poor ventilation increasing the amount of energy required to achieve thermal comfort. Dilapidated housing often has poorly maintained heating and/or cooling equipment as well as poor insulation. Aside from the health issues associated with thermal discomfort described in Chapter 3, poorly maintained gas heating can lead to carbon monoxide leaks. Carbon monoxide is a colourless and odourless gas that has the same mass as the O_2 molecule required for human respiration. Thus, inhalation of CO can lead to loss of consciousness and can prove fatal when exposure occurs for a long enough period of time.

Thus, performing essential maintenance and repairs on housing and associated appliances can be essential to health and safety of the residents. Good affordable housing policy should incorporate provisions to aid people to maintain their house in good condition, even when their monthly accommodation costs appear to be only a small percentage of the monthly income. This could include programmes offering subsidies to install better insulation, offering discounted maintenance on major electrical appliances and assistance in performing essential regular repairs that can prevent houses falling into dilapidated condition. Kumar et al. (2004) noted that clever housing design can go one step further and actually help prevent the spread of disease. As an example, they note that houses that include cross-ventilation, meshed doors and windows, and concrete roofs deter mosquitoes, which can prevent the spread of mosquito borne illnesses such as malaria. In low income countries malaria is one of the top ten causes of death according to the World Health Organisation (2018) and is particularly prevalent in Sub-Saharan Africa and South East Asia.

Urban versus rural issues

Affordable housing is often seen as less of a problem in rural areas compared to urban, because housing prices are generally lower. However, rural incomes are also generally lower, so affordable housing issues are actually comparable (Beer, 1998; Ziebarth, Prochaska-Cue, and Shrewsbury, 1997). In fact, in terms of adequacy rural housing is at higher risk due to the low population density. A dilapidated house in a busy urban area would be regularly seen by many people, increasing the chance of its poor condition being noticed. The building would thus more likely be reported, not just over concern for the house's residents, but the building could pose a danger to people living in the surrounding buildings. In a sparsely populated rural area, a house may rarely be seen by others and so the dilapidated conditions would go unnoticed. Morton et al. (2004) and Ziebarth et al. (1997) both note that in the U.S. the adequacy of housing in rural areas can be far worse than in urban areas. This was reportedly largely due to the increased likelihood that rural properties are owner-occupied and thus, are dilapidated as a result of the owner's inability to afford repairs and maintenance costs.

On the other hand, the close-knit community spirit of rural villages could also mean that dilapidated conditions in rural housing are more likely to be noticed by people who have genuine concern for the residents. Morton et al. (2004) explored this idea in rural U.S. and noted that the "sense of community" of a place had a direct bearing on the state of housing. In regions with a strong sense of community, houses were less likely to be dilapidated, independent of incomes, compared to other locations. In Cornwall, U.K. a young man lost his family's farmhouse and it was put up for auction. The local community banded together to help him regain his home by stopping anyone else from bidding during the auction, keeping the price to a minimum and ensuring the young farmer could afford to buy it back (Holidays in Cornwall, 2018). Thus, being a part of a caring community can enhance housing affordability in the absence of effective policy intervention. This further highlights the advantages of supporting the development of socio-economically diverse communities.

Overcrowding

Overcrowding is another form of housing inadequacy. The definition of overcrowding varies regionally but usually uses the measure of persons per room or persons per bedroom, often incorporating restrictions related to age and gender. For example, The Canadian National Occupancy Standard, which is also used in Australia (Australian Institute of Health and Welfare, 2018), defines overcrowding as:

A measure of appropriateness of housing that is sensitive to both household size and composition, specifically:

- No more than 2 people shall share a bedroom
- Parents or couples may share a bedroom
- Children under 5, either of the same sex or opposite sex, may share a bedroom
- Children under 18 of the same sex may share a bedroom
- A child aged 5–17 should not share a bedroom with a child under 5 of the opposite sex
- Single adults 18 and over and any unpaired children require a separate bedroom.

Less frequently the definition of overcrowding will be a measure of persons per unit area and in some cases, such as Wales, it incorporates a combination of these measures (Shelter Cymru, 2017).

In Australia, overcrowding is considered a form of homelessness as it results in the failure to provide a secure, stable, private and safe space. Living in overcrowded conditions affects everybody. It can impact sleeping patterns, making it harder for children to concentrate in school, or adults to concentrate at work. For children sharing their sleeping space with adults it exposes them to adult problems and limits their privacy and protection. Teenage children of different genders forced to share a room has been shown to increase the risk of sexual assault. Overcrowding can also hinder social development by making it impossible to entertain guests at home, which is an important part of many cultures (Council to Homeless Persons, 2018). A detailed study commissioned by the U.K. Office of the Deputy Prime Minister (2004) investigated the health impacts of overcrowding. This found evidence that the physical health of children living in overcrowded conditions is impacted and further that adult health impacts are experienced as a result of overcrowded living conditions during childhood.

Lerman and Reeder (1987) developed a new equation of affordability that included a measure of adequacy. The formula incorporates a value of the minimum cost of an adequate house, rather than simply basing it on the cost of the house. In this case an adequate house is defined as providing decent, safe and sanitary accommodation. Thus, households spending less than 30 per cent (or whatever the conventional definition of affordability entails) of their monthly income on a house will still fall under the definition of affordable if the quality of the house meets a certain standard. Similarly, a household choosing to spend a high proportion of their salary on their housing, but could easily afford a house of minimum adequacy will be removed from the definition.

Availability

Another approach to determining affordability involves assessing whether a household's income could afford the average price of a house in their area (Bramley and Watkins, 2009). The equation must specify the average house meeting a minimum standard to ensure adequacy and in this way will simultaneously exclude people on high incomes choosing to spend more on their house, while including low to medium income houses living in inadequate conditions. The equation is applicable to either mortgage or rental payments. However, for mortgage payments it is important to factor in likely repair and maintenance costs to be incurred, while for rental properties projections of likely fluctuations in rental prices must be considered. The potential problem with this formula is the assumption that houses at this mean price are readily available, which may not be the case, and brings us to the third component of the affordability equation: availability.

The U.K. provides a good example for this, as the mean house price in many rural regions falls comfortably within the affordability of even lower income residents. However, the number of properties available is less than the number of people looking for accommodation (Bramley and Watkins, 2009). The lack of available housing is attributed to the trend for second home purchases and strong opposition to new housing developments. People who purchase a rural property as a second home intend to use it for vacations, and so do not want development occurring in the region as it could potentially destroy the aesthetic appeal of a rural setting. In particular, the thought of affordable housing conjures up images of decrepit council estates full of drug users and criminals. While environmental concerns are often raised in opposition to new housing development Coombes (2009) argues that the real motivation for opposing new housing developments aimed at low-income households stems from wealthy elites wanting to keep undesirable types out of their area. The high proportion of second homes in these regions also brings up the argument of who is more entitled to have their say over development decisions affecting an area – second home owners who only spend a small amount of time in the area, or local residents who live there permanently. Unfortunately, homeowners tend to hold more sway over the local councils and more often than not get their way when it comes to blocking affordable housing development schemes.

Figure 4.3 shows a clear roadmap for legislative support on affordability solutions under two key provisions, home ownership and rental assistance. Whether the legislative support and government mechanism should promote home ownership or rental assistance to the needy community depends on a number of important factors. Some of the key factors are *land cost, building cost, time and quality, empowering community, growth in value, repair and maintenance, value for money, sustained expenditure* and *access to fundamental services*. Irrespective of whether the fund is public or private, if the property

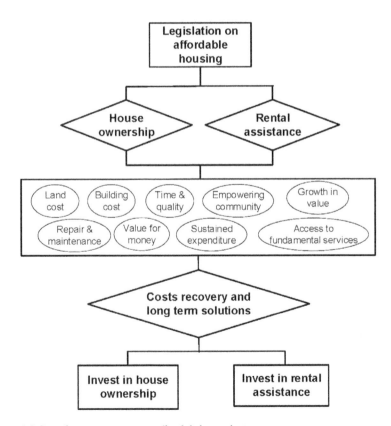

Figure 4.3 Legislative support on affordability solutions

development investment has no long-term growth potential, rental may be a better solution. Over time, home ownership is usually 20–30 times more expensive than a renting, thus government may not be able to justify financial support for the expansion of home ownership in rural communities.

The role of government

There are countless examples of government intervention into the provision of affordable housing. These broadly take either a comprehensive role, where the government oversees the whole project, or a supplementary role, where the government offers financial assistance, but other stakeholders implement the scheme (Lundqvist, 1986). These various policy instruments can be grouped into three broad categories: 1) expanding the affordable housing base; 2) provision of subsidies; 3) financial sustainability mechanisms (Haylen, 2015).

The first category includes comprehensive development schemes where the government procures the land, engages developers and oversees the building, and facilitates the allocation of houses to those in need. This approach was particularly prevalent in post-World War II Europe in trying to replace the many buildings that had been lost during the bombing. The U.K. is an example of this post-World War II government housing scheme, however, by the 1980s the costs to the government of maintaining this housing base became unfeasible and resulted in the government initiative known as the right to purchase (Whitehead, 1990). Families currently living in government-owned housing were offered the opportunity to purchase the property at a price well below market value. Once the house was purchased, the owners were free to sell it on at market value, thus the scheme was very beneficial for many low-income households. However, it also led to a severe reduction in the amount of accommodation available for people in need.

In India, the government takes a comprehensive role in the provision of affordable housing. The schemes are improving, as we will see in Chapter 7, however, in the past these schemes have faced many problems due to corruption. Corrupt government officials tended to grant development contracts to friends or relatives, sometimes at inflated prices. In other cases, corrupt developers have scammed the government. In one instance the government granted a developer an affordable housing contract, giving them ownership of the land on which the development was to occur (Mukhija, 2004). A short while into the scheme, the developer declared bankruptcy and sold off the remaining land to another developer. This second developer had no obligation to uphold the affordable housing contract and proceeded to build high-end homes which were sold to high income earners. It was rumoured that the second developer was a relative of the initial developer. In another case a developer took a commission to construct affordable housing, but built houses that were inadequate. Some were incomplete, others had appliances missing, or had not been connected to infrastructure

(Mukhija, 2004). After moving in, the residents began to experience problems with the shoddy buildings immediately, but the developers refused to perform any repairs, claiming they had fulfilled their contractual obligations and were not required to do anything else.

Another criticism of category 1) approaches is the appropriateness of the housing. While we have already described how the quality of government funded housing can be compromised, there is often no flexibility in the design of these houses. Therefore, the layout of the houses will not be able to cater to households with different cultural needs. In fact, it would be near impossible to provide a family with suitable housing without individually tailoring the design of each house. However, this would not be economically feasible for a comprehensive government funded affordable housing scheme.

Because of this, many consider a supplementary approach from the government more appropriate. In the U.S. in particular, there is strong opposition to the government's direct involvement in housing development, and enabling affordable housing through regulations is preferred. Category 2) schemes, where people receive subsidies to help with their accommodation costs, is an example of a supplementary approach. In many developing countries this subsidy comes in the form of tax relief for homeowners assisting with their monthly accommodation expenses (Lundqvist, 1986). It is important that the scheme is restricted to families in need. In some schemes, the benefit was offered indiscriminately, meaning even higher income earners received the tax break. This inadvertently inflated the price of housing by giving people a greater proportion of their income to spend on mortgage repayments. As a result the subsidy actually made it harder for lower income families to afford accommodation.

Another concern over the provision of subsidies is the longevity. In some regions it seems that once a household is found eligible to receive a housing subsidy, they will have it for life. The tenability of this situation has been questioned. It has been argued that the purpose of the scheme should be to enable people to improve their circumstances and thus it should only be a temporary measure. This comes back to the example described earlier, where the subsidy allowed the resident to live close to an employment opportunity. Once they start the job and begin earning, they will be able to pay off any debts and begin covering more of their own expenses. It is often also assumed that over time their wage will increase, due to inflation and potentially promotion depending on the type of job.

Rather than targeting residents, some government schemes target the construction industry, providing incentives for them to construct affordable houses. Various ways of implementing this approach exist. For example, when a developer applies for planning permission to develop in a highly sought after area, the permission may be granted with a caveat requiring a set number of the properties to be sold at a fixed percentage below market value. These properties would only be available to purchase for people that meet a certain eligibility criteria. While these can be effective solutions, they are only short term ones. The people that purchase the affordable property are often required to remain living in it for a fixed number of years, but after this time are free to sell it on at market value.

So far we have considered government funded affordable housing schemes, however, there are other approaches to housing provision. Non-governmental organisations such as Habitat for Humanity are dedicated to the provision of affordable housing. These organisations arrange teams of volunteers to work together with a family in need to help them build a house. By participating in the building process, the future residents have some control over the final layout of their home, and have the opportunity to develop new construction skills. This can give them the skills to perform some essential repairs and maintenance work in the future, enhancing affordability, and potentially increasing their employability. In Brazil, businesses work together with underserved communities to aid the provision of housing. Often these are funded through partnerships between governments and NGOs or other businesses, easing some of the financial burden on the government (Bowns and da Silva, 2011) These housing development schemes embrace the concept of "community practice" and facilitate the social, physical and economic development of underserved communities.

Cooperative housing schemes are an informal mechanism that can provide affordable accommodation for low income earners. The central idea behind these schemes is a group of low income earners applying for a joint loan. This allows them to apply for a low-interest rate mortgage, when they would only be eligible for a subprime mortgage. The group may then purchase one or several properties and live communally. Thus, they may jointly purchase an apartment block with a separate flat for each family, or they may purchase a large house with people occupying their own room. In some instances they will have to share some of the facilities such as a bathroom and kitchen. In others, residents eat together and share chores.

While this may not be an attractive prospect for some, it is a growing trend. In Australia, where many young people struggle to become eligible for a home loan, buying jointly can have various benefits. The mortgage will be paid off faster, minimising interest, they are improving their credit rating, and their future buying power will theoretically increase assuming the value of their property increases. The communal living in cooperative housing schemes can also have social benefits for some. For example, single parents may find that they can share childcare responsibilities, freeing up their time. Only children may enjoy having other children to play and study with.

A recurring theme throughout the literature on affordable housing is the importance of public consultation. Although a comprehensive approach to affordable housing provision may seem convenient, the design of the housing will never suit every household under their jurisdiction. Individual regions need tailored solutions to ensure the adequacy of their home (Ramamurthy, 1989). Ramamurthy further goes on to say that the construction industry should take some initiative to design affordable housing options without waiting for instructions from the government. Dorsey (1989) also notes the importance of focusing on the end user at all stages from design to construction. They note that poor communication between architects, engineers, construction workers and facility managers often hinders good quality affordable housing. If all these players could work together more effectively, with greater emphasis on the needs of the end users, better design and outcomes would result.

References

Acolin, A. (2018). Better location, better housing: Incorporating location into affordable housing loan programs. *Housing Finance International* (Autumn), 16–24.

Australian Institute of Health and Welfare. (2018). Housing assistance in Australia 2018. Retrieved from www.aihw.gov.au/reports/housing-assistance/housing-assistance-in-australia-2018/contents/overcrowding-and-underutilisation

Beer, A. (1998). Overcrowding, quality and affordability: Critical issues in non-metropolitan rental housing. *Rural Society*, 8(1), 5–15.

Bowns, C., and da Silva, C. P. C. (2011). Community practice, the millennium development goals and civil society measures in Brazil. *International Journal of Architectural Research*, 5(2), 7–23.

Bramley, G., and Watkins, D. (2009). Affordability and supply: The rural dimension. *Planning Practice and Research*, 24(2), 185–210.

Chetia, A. (2018). Is JAM enough to ensure financial inclusion of the rural economy? Paper presented at the 1st International Conference on Smart Villages and Rural Development (COSVARD), Guwahati, India.

Coombes, M. (2009). English rural housing market policy: Some inconvenient truths? *Planning Practice and Research*, 24(2), 211–231.

Council to Homeless Persons. (2018). No room to breathe: Why severe overcrowding is a form of homelessness. Retrieved from https://chp.org.au/no-room-to-breathe-why-severe-overcrowding-is-a-form-of-homelessness/

Dolbeare, C. N. (2001). Housing affordability: Challenge and context. *Cityscape: A Journal of Policy Development and Research*, 5(2), 111–130.

Dorsey, R. W. (1989). Integration of architectural and engineering skills. In O. Ural and L. D. Shen (eds), *Affordable housing: A challenge for civil engineers*. New York: American Society of Civil Engineers.

French, S., Leyshon, A., and Thrift, N. (2009). A very geographical crisis: The making and breaking of the 2007–2008 financial crisis. *Cambridge Journal of Regions, Economy and Society*, 2, 287–302.

Hatton, B. (2012). Portugal scraps rent control. *Multihousing Pro Magazine*. Retrieved from www.multihousingpro.com/article.php?AID=823

Haylen, A. (2015). Affordable rental housing: Current policies and options. Document 11, NSW Parliamentary Research Service.

Hewson, B. (2012). *Investment in affordable housing and housing microfinance in Africa*. New York: New Urban Finance Facility for Africa.

Heyford, S. C. (2019). What is a subprime mortgage? *Real Estate*. Retrieved from www.investopedia.com/ask/answers/07/subprime-mortgage.asp

Holidays in Cornwall. (2018). Farmers stand together in silence so a young man can buy back his family farmhouse. Retrieved from www.holidaysincornwall.com/farm ers-stand-in-silence/

Kumar, D. S., Krishna, D., Murty, U. S., and Sai, K. (2004). Impact of different housing structures on filarial transmission in rural areas of southern India. *Southeast Asian Journal of Tropical Medicine and Public Health*, 35(3), 587–590.

Kutty, N. K. (2005). A new measure of housing affordability: Estimates and analytical results. *Housing Policy Debate*, 16(1), 113–142.

Lerman, D. L., and Reeder, W. J. (1987). The affordability of adequate housing. *AREUEA Journal*, 15(4), 389–404.

Luffman, J. (2006). *Measuring housing affordability*. Perspectives 75-001-XIE, Statistics Canada.

Lundqvist, L. J. (1986). *Housing policy and equality*. Dover, NH: Croom-Helm.

Maclennan, D. and Williams, R. (1990) *Affordable housing in Britain and America*. York, U.K.: Joseph Roundtree Foundation.

Morton, L. W., Allen, B. L., and Li, T. (2004). Rural housing adequacy and civic structure. *Sociological Inquiry*, 74(4), 464–491.

Mueller, E. J., and Tighe, J. R. (2007). Making the case for affordable housing: Connecting housing with health and education outcomes. *Journal of Planning Literature*, 21(4), 371–385.

Mukhija, V. (2004). The contradictions in enabling private developers of affordable housing: A cautionary case from Ahmedabad, India. *Urban Studies*, 41(11), 2231–2244.

Nguyen, M. T. (2005). Does affordable housing detrimentally affect property values? A review of the literature. *Journal of Planning Literature*, 20(1), 15–26.

Office of the Deputy Prime Minister. (2004). *The impact of overcrowding on health and education: A review of the evidence and literature*. London: Office of the Deputy Prime Minister.

Paris, C. (2007). International perspectives on planning and affordable housing. *Housing Studies*, 22(1), 1–9.

Ramamurthy, K. N. (1989). Shelter for the homeless. In O. Ural and L. D. Shen (eds), *Affordable housing: A challenge for civil engineers*. New York: American Society of Civil Engineers.

Salesi, Jenny. (2019). Opportunities and challenges in Pacific housing. In *Pacific Peoples Housing Forum*. Auckland, New Zealand.

Shelter Cymru. (2017). Overcrowding. Retrieved from https://sheltercymru.org.uk/ get-advice/repairs-and-bad-conditions/overcrowding/

Tighe, R. (2010). Public opinion and affordable housing: A review of the literature. *Journal of Planning Literature*. Retrieved from https://journals.sagepub.com/doi/10. 1177/0885412210379974

United Nations. (2019). Goal 11: Sustainable cities and communities. *Sustainable Development Goals*. Retrieved from www.unenvironment.org/explore-topics/susta inable-development-goals/why-do-sustainable-development-goals-matter/goal-11

Whitehead, C. M. E. (1990). Housing finance in the U.K. in the 1980s. In D. Maclennan and R. Williams (eds), *Affordable housing in Britain and America*. York, U.K.: Joseph Roundtree Foundation.

Wiggers, K. (2019). Pew: Smartphone penetration ranges from 24 per cent in India to 95 per cent in South Korea. Retrieved from https://venturebeat.com/2019/05/27/south-am erican-countries-denounce-decision-to-give-amazon-control-of-amazon-domain/

Wise, P. (2017) Lisbon stalls on rent and lease reform. *Financial Times online* accessed 16 June 2019. Retrieved from www.ft.com/content/7160c322-fe70-11e6-8d8e-a 5e3738f9ae4

World Health Organisation. (2018). The top 10 causes of death. Fact Sheets. Retrieved from www.who.int/news-room/fact-sheets/detail/the-top-10-causes-of-death

Ziebarth, A., Prochaska-Cue, K., and Shrewsbury, B. (1997). Growth and locational impacts for housing in small communities. *Rural Sociology*, 62(1), 111–125.

5 Housing affordability

A house in and of itself is not the only factor to consider in relation to affordability. The layout and location of the house is equally important to a household's cost of living and quality of life (Acolin, 2018). In many regions a seemingly economic solution to the development of affordable housing has been building a mass of units on land that is cheap either because it is inconveniently located or poor quality due to previous development. We will be discussing issues related to accessing resources including land tenure in Chapter 6, but in this chapter, we will look at some of the ways in which living in an inconvenient location can negatively impact both affordability and quality of life of the residents. Conversely, the location could provide an opportunity if located close to employment opportunities and essential services. As we saw in Chapter 3, the layout of a house can also impact on a household's lifestyle. We will build on this idea in this chapter, focusing on how it impacts affordability.

Perhaps the most important factor to consider is accessibility to transport. People often need to perform daily activities outside their home including work, attending school, and meeting up with friends and family. The ability to perform these activities is fundamental to a person's well-being and will rely on their ability to access the locations where these activities are performed. Therefore, the accessibility of transport required to reach these locations is an important consideration. The first section of this chapter will focus on issues related to transport access. Food is also essential to life and housing provision can impact this in two ways: Firstly, in many rural communities of developing economies the people are in the habit of growing their own food. Therefore, having access to arable land, or space to safely house livestock will be important to their continued well-being. Secondly, the method of preparing food and the ritual of eating food is an important part of many cultures. Therefore, affordable housing should also provide appropriate cooking facilities and eating spaces. The second and third sections of this chapter will describe these two issues of space in detail.

In Chapter 2 we saw that the community spirit of rural villages was one of the main advantages over life in an urban area. As Powers, Davis, and Loza (2000, 2) puts it, "housing is more than roofs and walls. It is about strengthening our communities, good homes go along with good health, stability, pride, caring

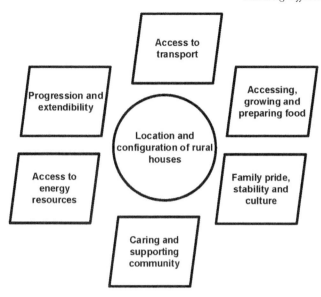

Figure 5.1 Location and configuration of rural housing

about our neighbourhood and having hope for the future." While affordable housing schemes should aim to nurture this sense of community, in the past they have ended up dividing communities. The fourth section of this chapter will discuss ways to ensure socio-economic diversity in rural communities and avoid segregation of different community members on the basis of their socio-economic status.

Infrastructure provision for rural communities is equally important to housing provision and will be discussed in the fifth section of this chapter. In fact, housing and infrastructure are intrinsically linked. Heating and cooking require energy resources; electricity is considered a basic human need and forms one of the United Nations Sustainable Development goals (United Nations, 2019b); clean water and sanitation facilities are important for good health. Yet, due to the structure of most governments, infrastructure and housing are developed separately. The sixth section of this chapter will look at integrating infrastructure and housing policy to promote developmental improvement in rural communities. Figure 5.1 depicts the location and configuration of rural housing in relation to the above considerations.

Transport

Limitations of poor transport infrastructure

Access to reliable transport is essential for obtaining and maintaining a good quality of life in many ways. For example, once a family becomes established in their new home, they are still vulnerable to losing their home if they cannot pay the rent or keep up mortgage repayments. The ability to cover the costs of any essential repairs and maintenance is also important to ensure the house remains adequate. Thus, people must be able to generate adequate income to continue to cover the cost of their housing. To do this, they must be able to access a place of work in a convenient and affordable way. Conversely, some people may perform income generating activities in their home. In particular, in rural areas of developing economies, many households are engaged in agricultural activities as their main source of income. However, to effectively generate income from these activities there must be a way to deliver the products to their consumers.

For children, the most effective way for them to break out of the poverty cycle has been found to be educational attainment (United Nations, 2019a). It has also been shown that improving children's educational attainment leads to long-term economic benefits for the society as a whole (Center for Global Development, 2006; Glewwe and Jacoby, 2004). In Chapter 3 we described the importance of incorporating a quiet space where children can study at home into housing design. Ensuring children have convenient access to a good school is also essential to their long-term quality of life. To be able to attend school it is essential that either a facility is located within walking distance of the home, or that there is access to public transport links allowing them to attend school in a convenient, safe and affordable way. Where children lack convenient access to their school, they may end up missing days of school, falling behind the other students and eventually dropping out (Kazeem and Musalia, 2017).

This is only one of the ways children's education attainment can be hindered by their house location. In concentrated areas of poverty, children may be living in close proximity to their school but the quality of the education received in these schools can sometimes by very poor. The poverty status of these children means many of them may have severe emotional problems and learning difficulties. Schools lack the resources to nurture these children, and instead will suspend them for minor offences such as using a mobile phone. Mora (2013) found that these suspensions lead to juvenile detention. Rather than being rehabilitated by this experience and supported to return to society, the majority of these children go on to the prison system once they become adults. This disturbing phenomenon, known as the "school to prison pipeline" illustrates one of the ways people become trapped in the poverty cycle due to the location of their housing. In more diverse communities, where better funded schools can cater to differently abled children, this pipeline can be avoided and children have a chance at a better life in the future.

Poor transport links can also limit access to healthcare facilities. Lucas (2011) notes that this is the case in rural South Africa and is a huge encumbrance for many people as HIV, which requires frequent clinical visits, is prolific in the region. In fact, it has been reported that people living in poverty are generally more susceptible to illness. Cook et al. (2002) interviewed people living in rural poverty in the U.S. and found none of the participants reported having good health. The children of the interviewees were also experiencing above average levels of poor health including delayed cognitive function, autism, ADHD and seizures. The trend of young people migrating to urban areas also means that underpopulated rural areas have a high proportion of vulnerable elderly residents. Providing more healthcare facilities in sparsely populated rural areas is not an economically viable solution, compared to providing improved transport links to existing facilities.

Access to transport is also essential to ensure people can interact with friends, family and the community as a whole. Socialising with family and friends provides a support base in multiple ways. It reduces the risk of developing the mental health problems that often result from living in poverty. In regions that have limited or no access to media such as television and the internet, interacting with other community members can be many people's only source of information. Thus, they are reliant on this interaction to find out about the availability of services, or even the news. In particular, it is important for people to be aware of local political news to ensure their participation in the democratic process. Where lack of transport leaves citizens unable to attend council meetings or elections they are prevented from participating in the decision-making processes used to determine policies that directly affect their everyday lives.

This will also hinder attainment of the United Nations Sustainable Development Goal 16, which aims to promote peaceful and inclusive societies (United Nations, 2019c). Some of the targets of this goal include "Develop effective, accountable and transparent institutions at all levels" and "Ensure responsive, inclusive, participatory and representative decision-making at all levels." Novel approaches to policy and planning decisions, such as the "community practice" programme in Brazil (Bowns and da Silva, 2011) specifically target underserved communities through increased community engagement. Such schemes are essential to ensuring local policy is not commandeered by local elites, leaving the needs of the poorest members of society unmet. Where people living in poverty are physically prevented from attending community meetings due to poor access to public transport near their housing, their interests will not be represented, and policies will continue to favour the wealthy.

Access to different modes of transport

In rural areas of developing economies most households do not own a car. Johnston (2007) studied transport use in rural Indonesia and found less than 10 per cent owned a motorised vehicle. A survey of households on Majuli Island, Assam, India conducted by the University of Melbourne found that less than one per cent of residents commute to work by car. Henseler and Maisonnave (2018) report that in South Africa less than 25 per cent of households own a car. Alveano-Aguerrebere et al. (2017) note that even in the more urbanised area of Morelia, Mexico only around 20 per cent of residents regularly travel by private vehicle.

In most of these studies walking was the most frequently used mode of transport. On Majuli Island walking was the most common form of transport used at 76 per cent, followed by bicycle at 13 per cent. Walking was also the most common mode of transport used in rural Indonesia at 90 per cent. In Morelia, Mexico the most common methods of transport used were public transport (40 per cent) and then walking (36 per cent). The availability of public transport in Morelia was likely to have been more reliable due to its higher level of urbanisation compared to the other regions mentioned.

While reliance on walking and cycling may be adequate for short journeys, using these to travel longer distances can become extremely impractical, often taking up a significant amount of time as well as being a physical burden. Lucas (2011) describes the situation in rural South Africa. An interview with a child found he had to choose between taking a bus to school or having lunch as his family could only afford one or the other. The walk to school was two hours each way, which in itself is a time-consuming and physically exhausting situation. However, even more concerning was that, due to the high levels of poverty in the region, the walk to school was rife with gangs, significantly risking the child's safety. The threat from the gangs was two-fold – firstly direct attack, robbing children of their lunch money and any other essentials they may be carrying, secondly gangs were aiming to recruit young people to aid their criminal activities. Thus, the walk was exposing the child to significant danger.

Thus, having convenient and reliable transport links is an important component for good quality of life. Yet, many affordable housing development projects fail to provide this. In the U.K. Gallent, Mace, and Tewdwr-Jones (2002) notes that people are often forced to move away from their home town to find accommodation they can afford. Their home town is often the location of their job, their extended family and their friends. Even if the new house is in a neighbouring village, the few miles they are required to travel to access all these things can be a significant loss to their daily life, if the transport links are infrequent and unreliable as is the case in many rural areas.

Two main hindrances exist in relation to the provision of reliable public transport to rural areas. Firstly, motorised vehicles often have difficulty accessing rural areas as the roads are either non-existent or badly deteriorated. Many studies have reported on the poor maintenance of roads in developing rural areas in many regions of the world. The Asian Development Bank (2003) notes the poor maintenance of roads in Kyrgyzstan, Laos, Pakistan, Philippines, Uzbekistan and Vietnam; Mwase (1996) describes the poor quality of roads in Tanzania; Pinard, Newport, and Rijn (2016) describes the deteriorated state of roads in Africa in general. The University of Melbourne during the survey of Majuli Island residents noted the poor quality of roads in rural India, particularly on the island. During periods of extreme weather, these roads can become impassable, essentially trapping people in their village until the weather subsides. These poor quality roads can also hinder the income of agricultural households. Produce will be bounced around during transit to market places, which will damage its quality and reduce the potential sale price (Steyn et al., 2015).

This widespread deterioration is attributed to lack of foresight during the road infrastructure planning process. While many governments happily invest in building new roads due to the publicity this can bring, they fail to budget for regular maintenance as this is less visually impressive (Asian Development Bank, 2003). Thus, once built roads are left to fall into disrepair, especially in rural areas where the roads are used infrequently and by uninfluential people. Based on the field data collected from Majuli in Assam, an alternative model for maintenance of rural roads with a community-centric approach has been proposed (Doloi, Green and Donovan, 2019).

The second hindrance is the provision of public buses or other forms of transport. Due to the low level of use, public transport in rural areas is given low priority and where services are provided they are often infrequent, unreliable and unsafe. On top of this they can also be prohibitively expensive. Johnston (2007) for example noted that in Indonesia for many people the nearest bus stop could be several kilometres away, often requiring the use of a motorcycle taxi. We have already noted details of South Africa from Lucas (2011) where the use of a bus often came at the expense of food. Technology could provide a potential solution, as noted by Rönkkö et al. (2017). While it is true that regular bus services may be uneconomical due to the low level of use, a demand response service could be provided instead. People would use an app to request when they required a bus journey say the night before, and then the bus would collect all the people that had booked. This would also improve the local government's understanding of the frequency of use of buses and most popular routes. A similar service was described in Canada for aiding elderly people in remote rural communities in accessing healthcare services (Ryser and Halseth, 2012).

Access to technology could be a limiting factor to the implementation of this approach, however mobile phone penetration is reportedly becoming fairly ubiquitous even in many developing regions (Statista, 2019). In fact, in some areas it has been noted that access to a phone is not as much of a hindrance as access to electricity to charge the phone. The World Bank reports that households living in poverty are more likely to have access to a mobile phone than clean water and sanitary toilets (World Bank, 2016). The other way to get around the poor transport links issue is ensuring people have access to housing close to their work. This is likely to involve the implementation of affordable houses in a more affluent area. The pros and cons of this approach will be discussed in the Community section below.

Sustenance

In Chapter 4 we saw how the cost of a house can force people to cut back on other basic essentials including food. This is only one of the ways in which housing can impact on a family's ability to feed itself. People living in rural communities often rely on their own produce for sustenance. Many people simply cannot afford to purchase all of the food required for their basic sustenance (Hunger Notes, 2018). Others, as we saw in the Transport section, may find it difficult or impossible to physically access markets due to poor roads, particularly during periods of bad weather when roads can become impassable for days at a time. The United Nations (2015) reports that globally more than one billion people are employed in agriculture, highlighting the necessity for many people to have access to land as well as a building as part of their housing.

On Majuli Island, the University of Melbourne found that over 80 per cent of households were growing rice, but that only 4 per cent of these were for income generation. Therefore, around 78 per cent are relying on their rice growing for their own sustenance. If these people were provided with a house that did not have sufficient land, they would be incurring costs of purchasing food, as well as the associated transport costs required to reach the market place. Thus, this could be off-setting any savings of living in cheaper accommodation. Not only the size, but the soil quality and water availability of provided land must also be considered to ensure it will support adequate crop growth. Affordable housing developments often occur on brown-field sites, that is, locations that were previously sites of industrial use. While it is understandable that this practice can prevent the conversion of green-field sites, helping to preserve biodiversity and control greenhouse gas emissions, the soil in many of these locations may be contaminated. These contaminants may accumulate in foods and be ingested by the residents, potentially leading to health problems. It will also inhibit the potential for food produced here to be sold. Instead, if good quality land can be made available, it may improve food production rates, improving food security and potentially leading to increased income generation.

It may not always be feasible to provide substantial outdoor space on good quality soils for every house in a development project. In these instances, supporting the development of a communal gardening area where community members work together to grow food and share the produce could provide a good option. Lautenschlager and Smith (2007) studied a community garden project in Minneapolis and found that apart from providing a source of nutrition, the project also cultivated better community spirit. In particular, children who participated in the scheme displayed better social skills than those who hadn't. Other studies have noted that gardening can improve social interaction. A recent movement in the U.S. to convert the typical green lawns found in many suburban homes into sites of food production has found increased interaction between neighbours. The combination of spending more time out the front of their house, thus increasing the chances of residents encountering their neighbours, and the features of the garden giving them something to discuss (Zevnik, 2012) has cultivated better community spirit along with produce.

In developed countries, despite most people having convenient access to cheap food, there is a reverse trend where people are turning back to growing their own food. This is driven by fears over the impact of intensive agricultural practices such as the excessive use of pesticides and fertilisers and genetic modification. Growing your own food has also been considered a potential way to cultivate healthier eating habits among children, in countries that are blighted with obesity epidemics. Lautenschlager and Smith (2007) interviewed children in Minneapolis some of whom were participating in the community garden project and some who weren't. While the study was too short term to develop an understanding of the long-term health choices of the children they noted that the participants were more aware of the health benefits of different kinds of foods and the environmental impact of food production. They also noted that participants were more willing to try new foods. Thus, the benefits to households having access to land to cultivate their own produce are many.

Provision of protein rich foods is also essential to health, especially the growth and development of children. There are many rural communities that rely on non-commercial fishing as their main source of protein (Cooke et al., 2018). There are risks to continued access to fishing for these people due to the changing face of rural areas. Firstly, they may be forced to move away from the water source, to access affordable housing. Secondly, fishing is a huge tourist attraction and can thus attract many visitors to a region, with the associated risks of losing housing access described in Chapter 2. The advent of tourists will not only impact the cost of housing, but will also increase competition for the limited stocks of fish. Natural fish stocks are usually managed by a licensing scheme. Poor locals, who are reliant on the fish for sustenance, may not be able to afford the licence, while tourists who are fishing for recreation can, thus this system may worsen the situation for local anglers.

The increased waste and pollution that tourism brings also poses a risk to local fishing communities. In many rural areas there are no formal solid waste management or sewage systems, and waste is flushed into local waterways untreated. Where the population is relatively small, the impacts of this are insignificant as they are dispersed into the water system. However, the additional stress of tourist facilities could create significant contamination making the fish dangerous to consume. While it seems obvious that any tourist development would want to preserve the local waterways, especially as they form the crux of the tourist attraction, there are countless examples of this not being the case (Anctil and Blanc, 2016). Developers are often interested in generating a quick return on their investment, and will thus build things cheaply and quickly, showing little concern for the long-term sustainability of the development.

The use of aquaculture, that is, fish farms, has been suggested as a potential alternative to reliance on fishing from natural waterways. Troell et al. (2014) investigated whether increasing reliance on aquaculture could play a part in food security. They found that if we let our natural waterways become depleted and come to rely on fish farms there is a huge risk that the feedstocks required to maintain fish-farming will be lost and thus fish farms will become useless. Improving the management of our natural fish stocks was considered preferable. Ensuring fish stocks are managed in a way that does not disadvantage poor rural residents who rely on this for their sustenance should be prioritised.

Cooking facilities

In this book we consider adherence to the three pillars of sustainability an essential component of adequate housing. While it is a well-established fact that food is essential for human life, and indeed zero hunger is the second of the UN Sustainable Development Goals, food and cooking provide much more than just sustenance. In nearly all cultures cooking and eating are not just a means to an end, they form important cultural rituals (Murcott, 1982). Bach-Faig et al. (2011) for example noted that the United Nations Education, Scientific and Cultural Organisation (UNESCO) recognised the Mediterranean diet as an Intangible Cultural Heritage of Humanity in 2010. For remote rural communities in developing economies, these rituals are essential to the preservation of their way of life. Therefore, when designing appropriate housing, incorporating features that facilitate these rituals are essential for a development to be considered socially sustainable.

In India, as well as many other developing countries, open biomass burning stoves are still used for cooking in many rural communities. There are some benefits to this over the use of fossil fuels. It is more economical, as most families use dung and crop residues as fuels for the open biomass burning. As these families are engaged in agricultural activities access to these fuels is often abundant and free, although it should be noted that sometimes collection of biomass fuels can be an arduous task involving walking long distances. Therefore, this process both provides a free source of fuel and prevents waste. However, the open burning of biomass indoors has the potential to cause negative health impacts particularly in relation to smoke inhalation (Smith, Aggarwal, and Dave, 1983; Laxmi et al., 2003; Joon, Chandra, and Bhattacharya, 2009). Concerns were especially significant for more vulnerable household members such as children and the elderly or people with lung conditions such as asthma. Joon et al. (2009) in particular visited a rural house during cooking and found that the smoke in the house was so thick they could not remain indoors. Smith et al. (1983) noted, however, that the stoves were generally located on the floor, so the cook was seated on the ground during the cooking process. This minimised their exposure to the smoke, which tends to accumulate closer to the ceiling.

In response to these concerns the government has attempted to transition these communities to fossil fuel based cooking practices. Laxmi et al. (2003) considered replacing biomass burning with kerosene stoves. However, they found that in the region kerosene is expensive and only available in limited supply. Therefore, many households would not be able to afford this and for those that could, it would not be available to use every day due to the short supply in the region. It should also be noted that more recently, the burning of kerosene indoors has been linked to negative health impacts included some lung cancers (Lam et al., 2012).

In another initiative, the government provided electric hobs to households (Joon et al., 2009). Subsequent assessment of the project's success found most people were still using biomass burning for at least some of their cooking needs. When asked why they said because the taste of the food was not the same when cooked on the electric hob. This failure to consider the cooking rituals of the local residents wasted funding on a project that failed to deter people from using hazardous open biomass burning methods.

It should also be noted that encouraging the transition from the use of biomass to fossil fuels does not comply with sustainability principles as fossil fuels are more expensive, more difficult to obtain and generate non-biogenic greenhouse gas emissions. Instead more appropriate ways to reduce the hazards of biomass burning have been suggested. A simple solution is to locate the stove in an outdoor area such as a veranda or courtyard. In fact, the practice of locating a stove in an outdoor area is already common in many vernacular houses in warmer climates. In colder or temperate regions this would not be appropriate as the cooking facility often doubles up as the main source of indoor heating. In these houses, installing stoves with a flue directing smoke outdoors would be the simplest solution, although it is important that the people living in these households are educated in how to prevent them from becoming clogged, otherwise they will stop working (Smith et al., 1983).

A more complex solution would be supporting the transition from open biomass burning to a more sophisticated system such as an anaerobic digester (AD) or gasifier. These systems still use biomass as a fuel source but in a more efficient and smoke-free way. Installing a large-scale anaerobic digester that could serve the whole village could have multiple benefits. For example, households could take turns providing fuel, reducing the burden of fuel collection from a daily to a weekly or even less frequent task depending on the size of the village. The difficulty with AD systems is due to their complex technology they can be difficult to maintain (Doloi, Green, and Donovan, 2019). It is essential that a local resident understands how to perform repairs and maintenance on the system otherwise it will soon become redundant.

In areas transitioning from an agricultural based to a tourism-based industry local traditional food has the potential to provide a big draw for tourists. Guerrero et al. (2010) explored tourists' attitudes and found that many visitors expect to try local foods as part of the tourism experience. There are two definitions of local food, one is food that is grown locally, and the other is food that is prepared locally from imported ingredients. Both of these definitions form niche tourism markets: the first can fall into the category of organic or fair trade foods appealing to tourists interested in sustainability issues, while the second could be considered gourmet foods and will appeal to tourists interested in unique cultural experiences.

Home cooking has been linked to better health outcomes (Mills et al., 2017) thus, providing good cooking facilities can also have beneficial health outcomes through supporting better nutrition. The two main hindrances to home cooking were found to be time constraints and access to fresh produce. This ties together all the issues of affordability considered so far. Poor transport access may mean people are spending hours commuting to and from work, leaving little time for cooking. Limited access to markets or lack of land available for growing crops can also prevent people from being able to prepare healthy meals at home. This highlights the importance of taking the location of a house into consideration before calling it affordable.

Community

In Chapter 2 we described how the sense of community found in rural villages is one of the main attractions making urban dwellers long to move to these areas. This sense of community is cultivated by socio-economic diversity. We have already described in Chapter 4 how concentrated areas of poverty, such as the housing projects in the U.S., lead to crime and drug abuse (Mueller and Tighe, 2007). Families whose economic situation forces them to live in these communities face two scenarios. They either end up getting caught up in the lifestyle, which leads to their children becoming trapped in the poverty cycle and eventually becoming involved in criminal activity as adults. Or else they may find themselves in states of chronic mobility (Cook et al., 2002) where they are forced to move to a new house multiple times per year.

Moving to avoid a dangerous neighbourhood is only one trigger of chronic mobility. Many people find themselves constantly moving to find employment opportunities. There are two dimensions to this. Firstly, many people lose their job due to frequent lateness or absence. This can be the result of unreliable public transport services making reaching a work place at the same time each day near impossible. For parents, especially single parents of young children, the necessity to transport their children to school before heading to work can exacerbate this problem, especially when the school and their place of work are far apart. We have already noted that people living in poverty are more susceptible to illness, and may therefore need to take frequent time off work. Again, having children exacerbates the situation, as they will also need to take time off when their children are ill. Schools have many vacation days compared to work places, however, parents living in poverty are unlikely to be able to afford childcare services and may be forced to take time off work to care for their children on vacation days.

One of the biggest problems with chronic mobility is that people never get to know their neighbours and develop supportive relationships that lead to a strong community. Living in a close-knit community can help relieve this situation as friends or relatives will be on hand to cover childcare. In some communities parents take turns transporting groups of children to and from school and can share the childcare burden during school vacations. As a result, parents can significantly reduce absences and lateness and improving their chances of maintaining a job.

Secondly, due to the high crime rates associated with concentrated areas of poverty, businesses will avoid establishing their premises in these areas. Businesses that already exist, and the employment opportunities they provide, may end up closing down or moving to a safer neighbourhood. Thus, over time, job opportunities in these areas will diminish, again forcing people to move to find work. Finally, the accommodation itself can be a trigger of chronic mobility. For people with mortgages, inability to make repayments will see the bank foreclose on the property. For people in rental properties, depending on local policy, rents could be increased at any time, making the property beyond their affordability, or the building owner may sell the property, again forcing them to move.

People who end up in affordable housing programmes can sometimes end up stuck in these systems for life. An initiative in the U.S. in the 1990s attempted to break this cycle by providing rental assistance for people close to an employment opportunity (Cook et al., 2002). The idea was that the house location would allow the person to hold down a steady job, and over time, their position would stabilise and they would no longer require assistance to afford their accommodation costs. Unfortunately, the programme was not as successful as intended as the jobs were minimum wage, and even working full time hours were not providing enough income to cover the costs of living. Cook (2002) calculated that at the income rate on offer the person would need to work 70 hours a week just to cover their rent. For parents, the situation was impossible as they had no people around to assist with childcare. Thus, living as part of a community, where friends and family are close by to help care for children may be more important than living in close proximity to work.

So far we have considered the benefits of living in a socio-economically diverse community to the low-income earners; however, the benefits go both ways. The broader community can also be positively impacted by having housing targeted to low income earners. For example, small businesses, particularly in the services industry, who need staff but cannot afford high wages, struggle to source staff locally in affluent areas (Gabriel et al., 2005). As a result, staff will be travelling into work from further away, and may face difficulties such as arriving on time as discussed previously. Where transport is expensive, the commute to work may not even be worthwhile. Thus, businesses will suffer with unreliable staff, and frequent staff turnover.

It should also be noted that the ability to afford a house in a practical location is not always limited to the lowest income earners. While government housing schemes sometimes provide accommodation for low-income earners in desirable areas, people on relatively high medium incomes are unable to afford accommodation close to their places of work (Haylen, 2015; Kutty, 2005). In this situation people are either forced to endure a long commute, or cut back on other basic expenses such as food and clothing, to cover their accommodation costs. This situation is known as housing induced poverty and has been noted in Australia and the U.S. Affordable housing schemes should not simply be a case of targeting the lowest income earners, but rather ensuring a range of accommodation is available in different communities, creating healthier, more diverse neighbourhoods.

Thus, providing affordable houses in a socio-economically diverse community benefits everybody and could give people a real chance at breaking the poverty cycle and attaining development goals.

Infrastructure

Energy

As described in Chapter 3 thermal comfort is essential to good health and good health is essential to thriving communities. People in poor health will struggle to work, and so struggle to generate adequate income to cover their basic necessities. Children suffering from chronic poor health suffer from delayed cognitive development and struggle to reach their full potential. Thus, the ability to afford adequate housing includes not only the cost of rent or mortgage repayments, but also the cost of energy required to retain thermal comfort. Sener (1989) notes the importance of including the costs of heating into affordability calculations for houses in cold regions. To keep operational costs to a minimum, affordable houses should be designed with climate appropriate features. While developers may claim that this is likely to increase the construction costs of a house, the decreases in potential energy costs will more than make up for this over time.

In Chapter 3 we saw how many traditional vernacular building styles were adapted to suit their climate. These houses were built from locally sourced materials and labour, showing that incorporating energy efficiency into building design need not be overly expensive. Maximising natural indoor lighting should also be prioritised particularly where people are working inside during the day. Unfortunately, these features are often overlooked in affordable housing schemes, with many preferring to use a simple modern building design, irrespective of location.

While architectural features can go some way to maintaining thermal comfort it is unlikely this will be able to continue indefinitely as increasing global average temperatures are seeing more extreme weather. Heracleous and Michael (2018) for example modelled the thermal conditions of school buildings in Cyprus at projected increased temperatures in the region to 2050. Currently, the buildings maintain good thermal comfort despite the high temperatures experienced during the summer because of the vernacular design features. However, as temperatures increase in the near future, this level of comfort will become more of a challenge to maintain. Therefore, provision of energy will be essential. For colder climates, biomass burning for both heating and cooking is widely practised. While this is a sustainable option, it is important that houses are designed with appropriate ventilation to avoid the potential health risks associated with smoke inhalation. In warmer climates, electricity access will be essential for any cooling systems, whether fans, evaporative coolers or air conditioning.

Access to electricity is one of the UN Sustainable Development Goals (Goal 7: Affordable and Clean Energy for All) (United Nations, 2019b). The UN estimates that one in five people worldwide lack access to electricity, and this deprivation is more common in rural areas. Developing rural areas that are connected to national grid electricity infrastructure are prone to frequent blackouts and brownouts. Governments favour provision of electricity to urban areas and industrial plants, therefore when demand exceeds supply residential properties in rural villages are often the first to be cut off. Giving rural communities electricity independence through the provision of micro-plants could be a more workable solution. Depending on local resources this could take the form of hydroelectricity if near running water, anaerobic digestion for agricultural communities and so on. Electricity independence could also be incorporated into the house itself through provision of a solar cell or wind turbine. It is important to ensure that if these are provided they are placed in such a way that very little interference with them is required. For example, a widespread programme in India saw households provided with a solar cell (Barman et al., 2017). However, these were provided loose and it was left up to the householders to place them. Many people failed to place them in direct sunlight and thus they did not fully charge up and failed to provide their quoted rate of electrical supply.

Water and sanitation

Access to clean drinking water and sanitation is essential to survival. Cleaning with water is very important for reducing disease and improving health.

For agricultural communities, water is also essential for crop and livestock production. However, housing in many rural communities is not connected to a water supply, or in other locations the water supply is contaminated and unsafe to drink. To be truly affordable housing should be located near to a source of fresh water, or have access to a water purification system. Similarly, basic provision of sanitation accessible throughout the year is a key requirement for affordable housing.

Technology

We have already mentioned that technology can be used to make public transport services in rural areas more viable and to provide farmers with access to important information about agricultural techniques. Technology can also bring many benefits to small businesses in rural areas. For example, websites such as Craigslist and Gumtree provide free advertising. Galloway, Sanders, and Deakins (2011) studied the impact of internet access on small businesses in rural areas of Scotland. The ability to reach customers and send and receive payments instantly was noted to have improved efficiency and customer relations. It also provided access to a greater range of suppliers allowing businesses to purchase stock at lower prices or with faster delivery rates. The final benefit noted was that it increased the potential customer base. However, it was noted that the majority of small businesses preferred to maintain a local customer base, especially those providing trade services, due to the necessity to travel to customers. Most businesses were only looking to maintain a liveable income, rather than aiming to expand their businesses.

Policy implications

We have seen in this chapter that for housing to be considered truly affordable and adequate it must be appropriately located with access to services and infrastructure. Despite this, most governments treat the provision of infrastructure separately to housing, as they are comprised of separate departments each taking care of their particular task independently of the rest of the government. In fact, in many regions different levels of government have differing responsibilities. For example, housing policy often comes from national or state level, while educational and health services are the responsibility of local governments. Often, local governments' main source of revenue is house taxes (Mueller and Tighe, 2007). Thus, if the majority of houses in their locality are low-income families, they will have to keep taxes lower to ensure affordability. As a consequence, the amount of funding they receive will be limited compared to a region with predominantly wealthier residents, impacting the level and quality of services. For example, educational facilities in these regions will have fewer resources and struggle to attract good teachers. Where pockets of poverty exist, house tax should be collected at the broader regional level and distributed evenly to allow equal services provision and give better opportunities for people living in poverty.

In rural Ireland, housing development policy is aimed at preserving the agricultural feel of the region by maintaining low housing density and preventing development on green spaces (Scott and Murray, 2009). In reality, agriculture is a declining industry in the region and many rural residents desire a property within a village to provide them with more convenient access to services. The locals feel this desire to preserve the countryside is aimed at pleasing tourists rather than the local community. Due to the preservation of low density housing, the provision of services such as public transport is inefficient, and as a consequence the services are infrequent and unreliable. The authors argue that if the housing department worked together with transport services they would see the inefficiencies their policies are creating and realise that higher density housing would be more appropriate for modern rural communities. It would also be beneficial if the departments consulted with the local community to get a better understanding their needs.

In some regions, citizens have given up on the government's ability to provide them with infrastructure. In Trinidad and Tobago, squatters in an informal settlement have taken the initiative to connect themselves to infrastructure (Ramamurthy, 1989). Despite their high level of poverty the members of the community have shown great innovation by sourcing building materials from waste, such as fashioning pipes to connect their homes to water infrastructure. Petcou and Petrescu (2018) also reports on communities fed up with waiting for governments to respond to their needs. In this case, members of the community came together and developed action plans, which could be carried out by the community, and took these to the local government for approval. This bottom-up approach to infrastructure development further illustrates how services provision could be more efficient if governments worked more closely with members of their communities.

Thus, issues of affordability go beyond the house as a building to incorporate the location and layout of the house, the infrastructure it is connected to and the community it is a part of. This illustrates that housing policy cannot effectively be implemented on its own, but should be a part of a more holistic approach to providing citizens with all their needs. Consultation with the end-users of these programmes is the best way to determine appropriate and effective policies.

References

Acolin, A. (2018). Better location, better housing: Incorporating location into affordable housing loan programs. *Housing Finance International* (Autumn), 16–24.

Alveano-Aguerrebere, I., Ayvar-Campos, F. J., Farvid, M., and Lusk, A. (2017). Bicycle facilities that address safety, crime, and economic development: Perceptions from Morelia, Mexico. *International Journal of Environmental Research and Public Health*, 15(1).

Anctil, A., and Blanc, D. L. (2016). An educational simulation tool for integrated coastal tourism development in developing countries. *Journal of Sustainable Tourism*, 24(5), 783–798.

Asian Development Bank. (2003). *Road funds and road maintenance: An Asian perspective*. Manila: Asian Development Bank.

Bach-Faig, A., Berry, E. M., Lairon, D., Reguant, J., Trichopoulou, A., Dernini, S., ... Majem, L. S. (2011). Mediterranean diet pyramid today. Science and cultural updates. *Public Health Nutrition*, 14(12A), 2274–2284.

Barman, M., Mahapatra, S., Palit, D., and Chaudhury, M. K. (2017). Performance and impact evaluation of solar home lighting systems on the rural livelihood in Assam, India. *Energy for Sustainable Development*, 38, 10–20.

Bowns, C., and da Silva, C. P. C. (2011). Community practice, the Millennium Development Goals and civil society measures in Brazil. *International Journal of Architectural Research*, 5(2), 7–23.

Center for Global Development. (2006). *Rich world, poor world: A guide to global development*. Washington D.C.: Center for Global Development.

Cook, C. C., Crull, S. R., Fletcher, C. N., Hinnant-Bernard, T., and Peterson, J. (2002). Meeting family housing needs: Experiences of rural women in the midst of welfare reform. *Journal of Family and Economic Issues*, 23(3), 285–316.

Cooke, S. J., Twardek, W. M., Lennox, R. J., Zolderdo, A. J., Bower, S. D., Gutowsky, L. F. G., … Beard, D. (2018). The nexus of fun and nutrition: Recreational fishing is also about food. *Fish and Fisheries*, 19, 201–224.

Doloi, H., Green, R., and Donovan, S. (2019). *Planning, housing and infrastructure for Smart Villages*. Abingdon, U.K.: Routledge.

Gabriel, M., Jacobs, K., Arthurson, K., Burke, T., and Yates, J. (2005). National Research Venture 3: Housing affordability for lower income Australians. In *Conceptualising and measuring the housing affordability problem*, Australian Housing and Urban Research Institute (eds). Melbourne: Australian Housing and Urban Research Institute.

Gallent, N., Mace, A., and Tewdwr-Jones, M. (2002). Delivering affordable housing through planning: Explaining variable policy usage across rural England and Wales. *Planning Practice and Research*, 17(4), 465–483.

Galloway, L., Sanders, J., and Deakins, D. (2011). Rural small firms use of the internet: From global to local. *Journal of Rural Studies*, 27, 254–262.

Glewwe, P., and Jacoby, H. G. (2004). Economic growth and the demand for education: is there a wealth effect? *Journal of Development Economics*, 74, 33–51.

Guerrero, L., Claret, A., Verbeke, W., Enderli, G., Zakowska-Biemans, S., Vanhonacker, F., … Hersleth, M. (2010). Perception of traditional food products in six European regions using free word association. *Food Quality and Preference*, 21, 225–233.

Haylen, A. (2015). *Affordable rental housing: Current policies and options*. NSW Parliamentary Research Service.

Henseler, M., and Maisonnave, H. (2018). Low world oil prices: A chance to reform fuel subsidies and promote public transport? A case study for South Africa. *Transportation Research Part A*, 108, 45–62.

Heracleous, C., and Michael, A. (2018). Assessment of overheating risk and the impact of natural ventilation in educational buildings of Southern Europe under current and future climatic conditions. *Energy*, 165, 1228–1239.

Hunger Notes. (2018). 2018 world hunger and poverty facts and statistics. Retrieved from www.worldhunger.org/world-hunger-and-poverty-facts-and-statistics/

Johnston, D. C. (2007). These roads were made for walking? The nature and use of public transport services in Garut Regency, West Java, Indonesia. *Singapore Journal of Tropical Geography*, 28, 171–187.

Joon, V., Chandra, A., and Bhattacharya, M. (2009). Household energy consumption pattern and socio-cultural dimensions associated with it: A case study of rural Haryana, India. *Biomass and Bioenergy*, 33, 1509–1512.

Kazeem, A., and Musalia, J. M. (2017). The implications of ethnicity, gender, urban-rural residence and socioeconomic status for progress through school among children in Nigeria. *Social Indicators Research*, 132, 861–884.

Kutty, N. K. (2005). A new measure of housing affordability: Estimates and analytical results. *Housing Policy Debate*, 16(1), 113–142.

Lam, N. L., Smith, K. R., Gauthier, A., and Bates, M. N. (2012). Kerosene: A review of household uses and their hazards in low- and middle-income countries. *Journal of Toxicology and Environmental Health B: Critical Reviews*, 15(6), 396–432.

Lautenschlager, L., and Smith, C. (2007). Beliefs, knowledge, and values held by inner-city youth about gardening, nutrition and cooking. *Agriculture and Human Values*, 24, 245–258.

Laxmi, V., Parikh, J., Karmakar, S., and Dabrase, P. (2003). Household energy, women's hardship and health impacts in rural Rajasthan, India: Need for sustainable energy solutions. *Energy for Sustainable Development*, 7(1), 50–68.

Lucas, K. (2011). Making the connections between transport disadvantage and the social exclusion of low income populations in the Tshwane Region of South Africa. *Journal of Transport Geography*, 19, 1320–1334.

Mills, S., White, M., Brown, H., Wrieden, W., Kwasnicka, D., Halligan, J., … Adams, J. (2017). Health and social determinants and outcomes of home cooking: A systematic review of observational studies. *Appetite*, 111, 116–134.

Mora, R. (2013). Feeding the school-to-prison pipeline: The convergence of neoliberalism, conservativism, and penal populism. *Journal of Educational Controversy*, 7(1) article 5.

Morton, L. W., Allen, B. L., and Li, T. (2004). Rural housing adequacy and civic structure. *Sociological Inquiry*, 74(4), 464–491.

Mueller, E. J., and Tighe, J. R. (2007). Making the case for affordable housing: Connecting housing with health and education outcomes. *Journal of Planning Literature*, 21(4), 371–385.

Murcott, A. (1982). The cultural significance of food and eating. *Proceedings of the Nutrition Society*, 14, 203–210.

Mwase, N. (1996). Developing an environment-friendly transport system in Tanzania: Some policy considerations. *Transport Reviews*, 16(2), 145–156.

Petcou, C., and Petrescu, D. (2018). Co-produced urban resilience: A framework for bottom-up regeneration. *Architectural Design*, 88(5), 58–65.

Pinard, M. I., Newport, S. J., and Rijn, J. V. (2016). Addressing the road maintenance challenge in Africa: What can we do to solve this continuing problem? Paper presented at the International Conference on Transport and Road Research, Mombasa, Kenya.

Powers, William, Davis, Charles B., and Loza, Moises. (2000). Why housing matters. *Rural Voices* 6(1), 2–3.

Ramamurthy, K. N. (1989). Shelter for the homeless. In O. Ural and L. D. Shen (eds), *Affordable housing: A challenge for civil engineers*. New York: American Society of Civil Engineers.

Rönkkö, E., Luusua, A., Aarrevaara, E., Herneoja, A., and Muilu, T. (2017). New resource-wise planning strategies for smart urban-rural development in Finland. *Systems*, 5(10), 12.

Ryser, L., and Halseth, G. (2012). Resolving mobility constraints impeding rural seniors' access to regionalised services. *Journal of Aging and Social Policy*, 24(3), 328–344.

Scott, M., and Murray, M. (2009). Housing rural communities: Connecting rural dwellings to rural development in Ireland. *Housing Studies*, 24(6), 755–774.

Sener, E. M. (1989). Energy affordability of housing in cold regions. In O. Ural and L. D. Shen (eds), *Affordable housing: A challenge for civil engineers*. New York: American Society of Civil Engineers.

Smith, K. R., Aggarwal, A. L., and Dave, R. M. (1983). Air pollution and rural biomass fuels in developing countries: A pilot village study in India and implications for research and policy. *Atmospheric Environment*, 17(11), 2343–2362.

Statista. (2019). Number of mobile phone users worldwide from 2015–2020 (in billions). Retrieved from www.statista.com/statistics/274774/forecast-of-mobile-phone-users-worldwide/

Steyn, W. J. V., Nokes, B., du Plessis, L., Agacer, R., Burmas, N., and Popescu, L. (2015). Evaluation of the effect of rural road condition on agricultural produce transportation. *Transportation Research Record*, 2473, 33–41.

Troell, M., Naylor, R. L., Metian, M., Beveridge, M., Tyedmers, P. H., Folke, C., ... de Zeeuw, A. (2014). Does aquaculture add resilience to the global food system? *Proceedings of the National Academy of Sciences of the United States of America*, 111(37), 13257–13263.

United Nations. (2015). Sustainable Development Goals. Retrieved from www.un.org/sustainabledevelopment/

United Nations. (2019a). Goal 4: Quality education. *Sustainable Development Goals*. Retrieved from www.un.org/sustainabledevelopment/education/

United Nations. (2019b). Goal 7: Affordable and clean energy. *Sustainable Development Goals*. Retrieved from www.un.org/sustainabledevelopment/energy/

United Nations. (2019c). Goal 16: Promote just, peaceful and inclusive societies. *Sustainable Development Goals*. Retrieved from www.un.org/sustainabledevelopment/peace-justice/

World Bank. (2016). *World Development Report 2016: Digital dividends*. Washington D.C.: World Bank.

Zevnik, L. (2012). Expression through growing food and cooking: The craft consumption of food. *Journal for General Social Issues (Društvena istraživanja)*, 3(117), 753–769.

6 Materials and resources in construction of affordable houses

In previous chapters we noted that one of the biggest contributing factors to the lack of affordable housing is the fact that there are simply fewer houses in the current stock of buildings than there are people. This phenomenon is not unique to a particular region but exists everywhere, from poor rural communities in developing economies to urban agglomerations in developed regions. There is also an emerging trend of family units demanding their own housing, rather than sharing with relatives as had been done traditionally. This trend is illustrated in declines in the average number of people per household, seen almost everywhere according to the United Nations (2017). In Australia, the average number of people per household has declined from 4.5 in 1911 to 2.6 in 2016 (Australian Government, 2019). In the U.S. average household size decreased from 3.3 to 2.7 between 1960 and 2018 (Statista, 2019b). In China it dropped from 3.5 in 1990 to 2.9 in 2011, then rose slightly to 3.2 in 2017 (Statista, 2019a).

Providing more housing involves two major costs: land to build on and construction. The first section of this chapter will discuss issues relating to land availability and ownership.

Construction costs include both building materials and labour. In many regions this can form the most significant part of the overall cost. Expensive modern building materials such as brick and concrete are assumed superior to cheap, locally sourced renewable materials. The use of cheaper building materials is considered a compromise on quality. However, the existence of ancient vernacular structures contradicts this belief. In the second section we look at whether cheaper building materials can produce houses with equivalent or even superior quality to modern building materials. In the third section we look a lower cost modern building materials and techniques and again consider whether these can provide high quality structures. Even where the costs of land and construction are minimised, many low-income families will still need some assistance with their accommodation costs. Thus, at least part of the cost burden will fall to the government. In the final section of this chapter we will look at two broad methods for governments to assist with housing provision for low-income families. It is obvious that the need to provide more housing is not a static issue, but one that will continue to increase in line with population growth. Therefore, any effective solution to housing shortages must look beyond the current situation to provide a long-term solution for continuous growth. Figure 6.1 shows a process flow involved in construction of affordable houses.

Land availability

Land in rural areas is generally cheaper than urban areas due to the lower demand for development. In fact, it can seem that compared to densely populated urban areas, rural communities appear to have access to endless tracts of vacant land. However, there are competing uses for land in rural areas, meaning much of it is not available for development. The increasing population not only requires more housing but also more food. Therefore, sufficient arable land for growing crops and grazing livestock must be retained. The conservation of biodiversity and the associated ecosystem services we rely on for survival, as discussed in Chapter 2, means it is important that large tracts of land remain undeveloped.

Even where land is available, attaining ownership rights of land can be complicated by different political climates. In some regions, ordinary citizens are free to purchase a piece of land and own it outright. This means that they can pass the land on to their children ensuring some form of financial security for their future. It also means they can sell the land and use the money to buy another piece of land if they decide to move. In other regions, the government owns the land, and although citizens can rent the land long term and are free to build on it, they do not own it outright and cannot pass it on to other members of their family or sell it and use the money to move.

India provides an interesting case study in relation to land ownership. During the British occupation, large tracts of rural land were owned by a single family. Peasants were under the employ of these landowners, and

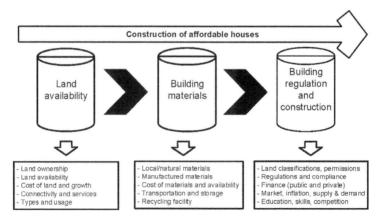

Figure 6.1 Construction of affordable houses

although they were free to live and work on the land, they never had any formal ownership. After independence, in a drive to reduce poverty land was to be distributed among these peasants (Walker, 2008). However, many of the existing landowners found loopholes in this policy, which saw them retain large tracts of their land. In fact Jeffrey (2000) notes the wealthy elites of some rural communities actually managed to gain more land and thus greater power in their community. Consequently, not only did wealthy landowners avoid having to give up their land to peasants but they were free to drive off any peasants currently living on their land (Walker, 2008). They would often later reemploy these people as low paid labour without the requirement to provide them with accommodation. Further attempts have been made since then to redistribute land but unfortunately have reportedly suffered the same failures. Thus, the rural poor who were the target of the schemes were worse off than they had been before. The main issue was that responsibility for implementing the scheme was at the local government level, and local governments were formed of the landowners. These local government officials were also in charge of recruiting the police force and used the police to enforce their personal agenda rather than the law. Any poorer members of the community who tried to speak out against the corruption were assaulted by the police.

In more recent times in India, there has been a drive to stamp out this corruption by requiring local government committees to contain enforced quotas of members from scheduled tribes (ST) and scheduled castes (SC). These ST and SC members are often in the employ of local elites and may be intimidated into agreeing with their agenda for fear of losing their jobs. However, there is some evidence that changes are slowly occurring (Tanabe, 2007).

Another more recent policy in India is the "housing for all" scheme. Details of the scheme will be discussed in more detail in Chapter 7, but basically the scheme provides funding for people to construct a house. For people who own a piece of land the scheme works very well, but for the landless it is more difficult. The scheme technically includes provision for tenure to the landless, however, it is reportedly very rare for this to occur (Kumar, 2018). Even when landless families of rural India have sufficient funds to cover the cost of a piece of land large enough for their needs, two barriers to obtaining landownership stand in their way. Firstly, a normal family may be able to afford a small plot of land, with sufficient room to build a house. However, the majority of rural properties are large farms that are not subdivided for sale. Thus, they are too expensive for these families.

Secondly, the legal fees for processing the landownership paper work are unimaginably high, reportedly far exceeding the cost of the land. To get around this, a government official who worked in the department for housing proposed a solution where a group of landless families would purchase a farm together, allowing them to divide it up to build their individual homes (Kumar, 2018). As the generator of the idea worked in the government department he used his position to ensure the participants were not hassled by government officials and forced to pay many expensive fees for processing the paperwork. The scheme was difficult to get off the ground as many of the landless did not trust the government. The department is known for making trouble for citizens in relation to their land tenure. However, the scheme was eventually successful and highlights that while provision of funding is helpful, reforming the legal system to make it easier for people to gain land tenure could be a much more effective step towards provision of affordable housing for all.

Informal settlements are another grey area for land tenure. There are various dimensions to this. In some regions, especially countries that were colonised by Europeans, indigenous tribes may have occupied land for centuries, but with no formal paperwork proving their right to occupation. Thus, European settlers drove these people from their land and took over ownership. Due to the traditional lifestyle of these tribes, earning formal income is difficult, thus hindering their ability to purchase a home. In many places these indigenous communities also faced racial prejudice and were often refused the right to purchase a home even where they could afford it. In many places, some headway into reversing this situation has begun including Canada, New Zealand, U.S., Norway and Sweden (Dow and Gardiner-Garden, 1998). In Australia, in 1976 the Aboriginal Land Rights Act granted indigenous communities formal ownership of lands occupied by their ancestors (Central Land Council, n.d.). This gives the people full control over any development, residential or commercial, on this land and rights in relation to any assets the land may contain.

Since 1957 the United Nations has attempted various international treaties on granting formal recognition of the rights of indigenous communities in all countries to both ownership of their traditional lands and the right to live traditional lifestyles (International Labour Organization, 1957). Despite this, some indigenous tribes are still vulnerable, particularly where their land contains valuable assets. In Papua New Guinea, large foreign corporations were granted permission to develop forested areas, under the guise that they were helping the local community develop their agricultural industry (Cool Earth, 2015). The access to land was supposed to have been granted with the consent of the local community who were entitled to the land. In reality, no such consent was granted, and these foreign companies logged the forested area, took all the timber and left, without any evidence of agricultural development. Instead, the land was degraded by the process and thus could not support crop growth.

Thus, it is important to recognise the rights of indigenous tribes to formal ownership of lands traditionally occupied by their ancestors, including the right to develop on these lands and control over any assets the land may contain. This will give indigenous communities the freedom to live their lives according to their cultural traditions without fear of homelessness.

Other forms of informal settlements come about when families either move into an unoccupied building, often because it is in a dilapidated condition, or piece together a house on public land. These are sometimes known as squatter settlements, although in many cases the occupants are paying rent to someone, just with no formal paperwork. That is to say, the landlord may not have formal ownership of the property and the tenants may not have a formal tenancy agreement. Durand-Lasserve and Royston (2002) looked at the different approaches that governments have tried to rectify this situation. Some tried to forcibly move people on, others tried to development alternative housing and move people into this. However, these approaches were rarely successful as the alternative accommodation was often in an inconvenient location. The most economically viable solution was shown to be granting land tenure to the occupants, allowing them to remain.

As discussed in Chapter 4 adequacy is a core component of an affordable house. These informal settlements are often comprised of dilapidated buildings, have no infrastructure connections, and are overcrowded. Again, the most economical approach to rectifying this situation is offering financial assistance to these households for repairs rather than moving them into a new house. The sheer number of people in these informal settlements usually makes the creation of new housing and forcibly moving the whole group too cumbersome, aside from the fact that the new housing is likely to be in a less convenient location. Non-governmental organisations often step in to assist with upgrading informal housing. World Habitat (2017) discusses a programme loaning money to households to put in windows for light and natural ventilation. By reducing their electricity consumption costs related to running fans and electric lights households can easily repay the loans.

Thus, while affordability can be a genuine hindrance to accessing land for housing, in many instances legal processes and political circumstances can be even more of a hindrance. Thus, to provide land for affordable housing, the rights of rural dwellers, particularly indigenous tribes, to outright ownership of land should be recognised. Similarly, the legal process for purchasing land should be simplified. Legal processing fees should not exceed the cost of the land and the process should be more accessible.

Building materials

The acquisition of land is a very political issue, but financially speaking in rural areas of developing economies land can be relatively cheap. Kumar (2018) for example noted that the legal fees associated with buying land in the Andaman and Nicobar Islands of India exceeded the cost of the land itself. The most significant portion of the cost of a new house is more commonly the building materials and construction labour. Ademiluyi and Raji (2008) noted that in Nigeria building materials account for 55–65 per cent of construction costs. Modern building materials in particular such as fired bricks, concrete, glass and steel, are expensive to manufacture. They are usually manufactured at a central location far away from rural areas meaning substantial transport costs will also be incurred. The poor transport infrastructure prevalent in rural areas will exacerbate the costs of trans-porting these materials.

Using cheaper building materials is one obvious way to reduce the cost of construction. However, this is often considered as compromising on quality. There have been instances where this has occurred, for example Gabriel et al. (2005) noted that in the U.S. developers that take on affordable housing contracts, they aim to get the houses built and sold as quickly and cheaply as possible with little or no concern for the quality or longevity of the building. As a result, they tend to use the cheapest building materials. Similarly, Mukhija (2004) notes that in India the Pashwanath group who were con-tracted to build affordable housing by the government used poor quality materials and left houses in an incomplete state to cut costs.

However, the use of cheaper building materials does not necessarily need to mean a compromise on quality. In Chapter 3 we saw that vernacular archi-tecture has produced buildings from locally sourced, natural materials for many generations that are adequate and functional for their occupants. The age of some of these structures is a testament to their high quality, for exam-ple in Egypt there are traditional vaulted mudbrick buildings that have been standing for thousands of years. Many of these building materials are virtually free. Thus, reverting to vernacular building materials could lead to significant reductions in the overall construction costs of a house. There are two poten-tial hindrances to this. Firstly, the construction techniques used to build the vernacular houses were usually passed on through generations by word of mouth, and have often not been documented anywhere. In regions where vernacular architecture has been out of favour for several decades there may no longer be any residents who remember these construction techniques.

Fathy (1973), an Egyptian architect, attempted to recreate a vaulted mud-brick house without a timber frame, such as is typically found in rural areas of Egypt. He assembled a team of engineers and architects, and examined existing structures, but in their various attempts to recreate the building, they could not work out how to support the roof without a frame. He went in search of people who still remembered the traditional building techniques and asked them to complete his building. Watching them work he noted how they did not use any measuring equipment but simply seemed to feel their way around the building while putting together the vaulted structure. Tracking down people with knowledge of vernacular architecture techniques and documenting their construction methods will be an important exercise for architects to ensure these techniques are not lost for good.

The second hindrance to revitalising vernacular architecture is the local perception. These houses have come to symbolise low socio-economic status while modern concrete and brick structures symbolise wealth and affluence. An extreme example of this desire to emulate modern housing design can be seen in rural Mexico. People who migrate to the U.S. for work send money back home for their families to build extravagant luxury houses that resem-ble those they see in the U.S. Many of these houses include letterboxes at the front despite the fact that in this region of Mexico there is no door to door mail delivery service (Lopez, 2010). Thus, many residents of rural communities desire to show their status through the style of their housing and will be unwilling to revert to vernacular structures, unless they see them embraced in modern urban design.

While the aesthetic symbolism of vernacular architecture has a strong part to play in the continuing penetration of modern architecture into the rural parts of the world, there are genuine concerns over the durability and resilience of vernacular building materials and designs. A part of this is concern over the ability of vernacular structures to withstand extreme weather events, earthquakes and other natural phenomena. We all know the story of the three little pigs, where building a house out of straw or sticks saw it blown down while the brick house was the only one that could withstand the wolf's attack. This perception persists, and fears abound that a house made of vernacular materials such as bamboo or mud-bricks will blow down in the next big storm or wash away in the next monsoon. The question then is whether these beliefs are based in fact, or if vernacular building materials can provide durable, resilient houses. Another concern that came up during a conversation with an Indian student who had firsthand experience of living in a vernacular house was the ability to decorate the interior space. Creating an interior atmosphere that reflects your family's values is important to a lot of people for turning their house from a mere building into a home. However, this student noted that the walls of the vernacular house were too weak to withstand hooks and he was therefore not able to put up the pictures he wanted to decorate his room with. As a result, he never felt truly at home there. Here we consider the validity of these concerns in relation to two of the most commonly used vernacular building materials – mud-bricks and bamboo.

Mud-brick

Mud-brick is a vernacular material that has been used in many rural areas due to its abundant availability and simplicity of production. Mud-bricks are made by packing together damp earth and leaving them to dry in the sun. Thus, their production has virtually no energy use and only requires access to dirt. As dirt can be found virtually anywhere, mud-bricks would rarely need to be transported long distances to a construction site. Therefore, they have very little embedded energy and thus, virtually zero greenhouse gas emissions. When the building reaches the end of its operational life, the bricks can be broken back down into dirt and returned to the earth they came from, meaning no waste is created. When built in a vaulted structure as described above (Fathy, 1973), the mud-bricks do not need to be reinforced with a wood or steel frame the way fired bricks do. This seemingly innovative technique was borne out of necessity as such materials were not traditionally available in these regions.

Mud-bricks have been found to have excellent insulation properties. Chel and Tiwari (2009) measured the internal temperatures of a mudbrick structure during different seasons. They found that during the winter internal temperatures remained between 14 and 18°C, while external temperatures fluctuated between 6 and 18°C. In the summer internal temperatures were maintained between 24 and 28°C, despite the external temperature fluctuating between 26 and 40°C. Thus, the interior space of the house was able to maintain a relatively comfortable temperature year-round minimising the need for energy consuming heating and cooling appliances. This makes it an ideal building material in both warm and cold climates.

Other benefits to the use of mud-brick include that they are resilient to fire and do not attract vermin, making them practical for use in rural houses, where open indoor open biomass burning is commonly used for heating and cooking. However, two concerns over the use of mud-bricks came up during our research. Firstly, their ability to withstand moisture and secondly, their ability to support interior design features such as shelves and picture hooks.

Rainfall will cause the mudbricks to rapidly deteriorate and without frequent maintenance parts of the building could become unstable and collapse over time (Chel and Tiwari, 2009). As the bricks deteriorate, moisture can build up inside the house, damaging interior furnishings and causing health problems for residents. The necessity to perform frequent maintenance on mud-bricks is not ideal, as it would not be required with modern building materials such as fired-bricks and reinforces the perception that modern materials are superior.

One way to prevent this is to make a house with mud-brick walls, but a different material for the roof, such as thatch. The roof must overhang the walls to ensure they are protected from rain. However, even with this in place, heavy rains can sometimes come in at an angle and may still reach the wall, particularly in areas that experience frequent monsoons or hurricanes. Also, moisture at the ground level can seep into the bricks. Having a different roof material may also affect the insulating properties of the mud-brick house. Ramesh, Prakash, and Shukla (2012) for example showed that roof insulation is more beneficial than wall insulation especially for warmer climates.

Another option is treating the bricks to improve their durability. One method is to coat the exterior surface of the bricks with a waterproofing substance. There are various commercially available products such as Dr. Fixit (Dr. Fixit, 2019). Another is to mix a reinforcing material into the mud mixture prior to drying. The use of reinforcing materials can overcome the two concerns mentioned above – it can improve resilience to water damage and increase strength, so the bricks can withstand stresses. Reinforcing can be done with natural, locally sourced materials such as straw. However, covering the mud-bricks to prevent contact with moisture remains the most effective way to maintain the durability (Dowton, 2013).

Bamboo

Bamboo is another building material that has been commonly used in vernacular architecture for thousands of years. Bamboo has been found in architecture in many parts of the world including South America, Oceania and Asia. While many builders understood inherently that bamboo is a strong building material, it was not until the mid-1990s that researchers began conducting formal strength testing on bamboo allowing its properties to be rated. This is largely due to the shape of bamboo making it impossible to test in traditional civil engineering testing equipment. Researchers at Eindhoven University of Technology in Holland, headed by Jules Janssen, tested and rated the properties of bamboo (Roach, 1996). They found that a column of bamboo can withstand greater compression forces than concrete. Tensile strength tests found bamboo to have similar properties to steel. It was also found to be able to withstand shear forces, making it tolerant of earthquakes. The beneficial properties of bamboo come from the fact that it is a naturally composite material comprised of strong lignin fibres and weaker cellulose. Thus, it is not only strong but also lightweight and flexible.

In terms of embedded energy, bamboo is very environmentally sustainable compared to other building materials. Bamboo grows much faster than the trees typically harvested for their timber. While most trees take many years to reach maturity, bamboo can be harvested annually (Nath, Das, and Das, 2009). The processing of bamboo into a building material is also a simple and low-energy consuming process. Despite this, over-harvesting of bamboo is still a risk. The uses of bamboo are many and varied (perhaps most famously as the main food source of pandas). These competing uses mean the bamboo can be over-harvested or harvested before reaching maturity in which case it will not have had time to develop the strong properties that make it a good building material. Therefore, if the use of bamboo is to become more prolific bamboo plantations must also be carefully managed.

There are two potential risks to the longevity of buildings constructed with bamboo. Being an organic material bamboo provides a source of food for many pests who can quickly eradicate a building, therefore treating the bamboo with pesticides prior to construction is essential. When it gets wet bamboo can expand and will quickly rot. Therefore, components of a house made from bamboo must be protected from rain. Where they are exposed to water, they must be well ventilated to ensure they can dry quickly and completely to prevent rot from setting in.

The final downside of bamboo as a building material is its high flammability, making it unsuitable for use in multi-storey buildings or urban areas. In single-story freestanding rural houses, the risks are minimised, however, the common use of indoor open biomass burning in many rural villages is of concern. These risks must be minimised by careful management of fires or converting houses to different forms of energy for heating and cooking.

However, the biggest hindrance to the widespread use of bamboo as a building material is its association with peasants. Bamboo structures have come to symbolise low socio-economic status while concrete and brick structures have come to represent affluence and high social standing.

Combining old and modern

While we have shown that traditional building materials can have similar or in some cases superior qualities to modern materials there are still instances where traditional materials are not able to cope. In particular, we discussed how both mudbricks and bamboo can be damaged by exposure to moisture, making their suitability for use in humid regions or locations with periods of heavy rainfall concerning. Overcoming the reluctance of rural residents to live in a vernacular structure due to the association with peasants is also a hindrance to their continued use. The solution may be a combination of vernacular and modern building materials. This can help minimise the costs of building materials without compromising on quality or aesthetics.

Many researchers have explored this idea, even as far back as 1962 (Kumar, 1962). An interesting housing programme in India embraced this concept by using traditional building materials for the bulky parts of the house structure, while incorporating the use of stronger modern building materials at the most vulnerable points of the building such as the joints (Bijlani, 1982). The modern materials reinforced the buildings to ensure durability and resilience, while the use of traditional materials minimised costs and improved environmental performance. More recent examples include Dodo et al. (2014) who described ways to combine the best features of the vernacular architecture found in the Sukur kingdom of Nigeria with modern building design to optimise structures.

Researchers from the University of Melbourne identified a pre-engineered Light Gauge Steel (LGS) frame fitted out with vernacular building materials such as bamboo and grasses as a good compromise for houses in the North Eastern part of India. The region has seasonal monsoons and as a result many rural villages experience flooding, therefore the typical bamboo frames used in the vernacular structures rot and need to be replaced frequently. The LGS frame is resilient to water, yet still light-weight and flexible enough to cope with the frequent seismic activity that occurs in the region.

Traditionally, the major cost in a building project goes to foundation work. The requirement of heavy machinery and skilled labourers for building traditional pile or raft foundations in remote locations create inefficiencies in the building process. The LGS frame proposed can be used with a concrete-less steel-plate and micro-pile foundation due to its light weight. As the structural components are pre-engineered, the laying of the foundations does not require expertise, and so can be done by virtually anybody.

The use of this frame with concrete-less foundations is suitable for most soil types including low lying land, flood prone areas, river banks and hilly terrains.

The ease of constructing an LGS framed building on a concrete-less stilt foundation system was exhibited at the Burnley Campus of the University of Melbourne, Australia by construction of a shed incorporating the technology. The frame of the two square metre shed (see Figure 6.2) was designed and produced at a factory about 50km away from the project site. The frame was delivered by a small truck and weighed about 170kg. The concrete-less foundation system of steel plate and micro-pile assembly was supplied by another company and used to erect the LGS frame by a group of unskilled students as a teaching related project. Figure 6.3 and Figure 6.4 show the project as installed on the site.

The demonstration shed building above consists of a metal shed structure. However, the LGS frame can be fitted out with many different building materials (see Figure 6.5). The use of these frames has been proposed for North East India, where the walls could be fitted out with locally sourced natural materials found in vernacular structures of the region such as bamboo or grasses. Thus, the houses can fit in with the local culture and value. Figure 6.6 and Figure 6.7 show two finished models of the LGS frame with vernacular external coverings and fit-outs on slab and stilt foundations, respectively.

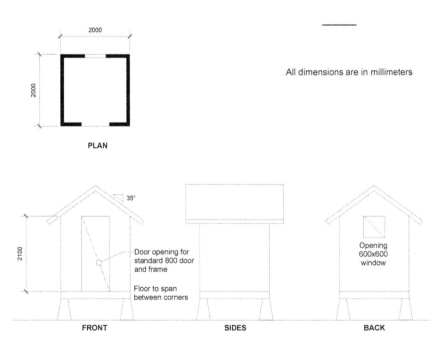

Figure 6.2 Plan of the two square metre shed

Figure 6.3 Complete LGS frame erected in less than four hours

Figure 6.4 Concrete-less steel plate and micro-pile assembly

Figure 6.5 Sectional view of the LGS framed housing model

Figure 6.6 LGS framed finished housing model on slab foundation

Figure 6.7 LGS framed finished housing model on stilt foundation

Cost-effective modern building

While so far, we have considered vernacular building materials a cheaper option to modern ones, there are ways to reduce the costs of modern building materials. Goswami et al. (2018) designed a wall made of Ferro cement panels, reinforced with chicken wire. Chicken wire is abundant in rural areas of India making it cheap and readily available. This novel material has lower embedded energy than fired-brick and was found to be 36 per cent cheaper. Yet testing showed it had comparable strength. The author built his own house out of these reinforced cement panels and produced photos of his beautifully decorated home, showing the walls were supporting shelving and artwork.

The foundations of a building often represent the most expensive part of the construction process. The necessary thickness of foundations is often decided at a regional level. As a result, locations with comparatively good soil structure for building on have the same foundation thickness as soils with very poor structure. Sazinski (1989) points out that by tailoring foundations to individual sites significant cost savings could be achieved by reducing the quantity of building materials required.

Plastics

Worldwide huge amounts of waste are being generated on a daily basis creating a crisis in management. While many of these materials have the potential to be recycled, this is not always possible due to contamination with other waste products during collection and a lack of suitable recycling facilities. The management of this waste creates a huge problem for all countries, however in developing countries the situation is somewhat worse due to restricted funding for waste management. For example, while land-filling is widely used in all regions of the world, the picture of this is very different. In developed countries landfills are carefully engineered structures that are completely sealed to prevent the release of toxic substances into the surrounding environment (for the most part, although there are often acci-dents). In developing countries however, landfilling is often simply dumping waste in a hole with little to no lining and cover. The waste penetrates the surrounding environment and the lack of security at these sites makes them accessible to the general public. Desperate people access these dumps, sort-ing through the garbage to look for anything of value they could potentially sell, putting their health and safety at serious risk.

The other main method of waste disposal is incineration. This can be an environmentally managed process, where the gas produced is converted to a fuel for energy generation and any toxic gaseous emissions are treated prior to release to prevent atmospheric pollution. However, it is also often the case that these pollutants are freely released into the environment creating a hazard for those living in the surrounding area. In some rural areas there are no formal waste management systems in place at all.

Many visionary engineers have started looking at ways to turn this waste from a nuisance into a useful resource. Waste from the construction indus-try forms around one third of total waste generated worldwide (Hoornweg and Bhada-Tata, 2012). The uniformity of this waste stream makes it ideal for recycling. Concrete can be ground down and remade into new concrete, timber waste if still in good condition can be reused, otherwise it can be mulched for composting or used as a fuel. Mixed waste, as long as it is inert and free from hazardous substances such as asbestos can be ground down and used in road surfacing. Using recycled construction waste can thus be a good way to minimise construction costs without compromising quality.

However, even more advanced engineers began thinking of ways to use products from the other waste streams that are more difficult to recycle and have little other options for management than landfilling or incineration. Plastics in particular are a huge problem for the waste management industry. The longevity of plastics is often lamented due to the hundreds of years they would take to degrade in a landfill. Their high carbon content means incin-eration would be contributing to GHG emissions. Of particular concern are lighter plastics such as single use carrier bags generated by the retail industry, which are difficult to recycle. Many recyclable plastics also end up in landfill because they become contaminated with food waste during the waste collec-tion process. However, this resistance to degradation can be turned into an advantage by turning plastic waste into a construction product. It is these types of waste that are finding a new purpose in the next generation of recy-cled building materials. Various organisations and companies have developed innovative solutions for incorporating this waste into a new product.

The simplest way to turn plastic waste in a construction material is to take a plastic bottle and fill it with lighter flexible plastics such as sandwich wrappers, single use carrier bags, candy bar wrappers (see for example WasteAid UK, 2019). These stuffed bottles act as a highly insulating, lightweight and robust building material. Many charitable organisations, including WasteAid UK, are encouraging the manufacture of these eco-bricks. In South Africa, the company Averda is supporting an education programme, targeting women living in poverty, in the upcycling of plastic waste into eco-bricks (Averda, 2019). The programme has the potential to bring many benefits to the region. Firstly, it encourages the women to collect plastic waste, removing it from the environment where it is a potential hazard, and upcycling it into a useful product. Secondly, the eco-bricks can be sold to the construction industry, creating a much-needed income generating opportunity. The company hopes that the support they are providing will encourage some of these women to turn their eco-bricks into small businesses. Thirdly, the programme is producing a low-cost building material to help with the housing shortage.

People wanting to manufacture a more sophisticated eco-brick can try the ByFusion blocker. This is a commercially available appliance that uses steam and compression to turn waste plastics into bricks. The environmentally friendly process can use any mixture of plastic types and sterilises the waste. Therefore, unlike most plastic recycling it does not require any sorting and is tolerant of contamination. There is also flexibility in the system that allows the shape of the blocks to vary to ensure they fit in with different building designs (ByFusion, 2019).

NevHouse is an Australia based company that produces wall panels from waste plastics known as composite recycled plastic panel (CRPP) (NevHouse, 2017). The product is made from low quality plastic waste that cannot be recycled in other ways. Usually, plastic waste recycling processes require a clean product that can only be obtained through either carefully segregated collection or energy intensive cleaning processes. However, CCRPs are tolerant of contamination allowing them to use contaminated plastics from mixed waste collections, and because the waste does not need to be sterilised the panels have a low embedded energy. Complete housing kits are produced that are easily assembled in just five days. While currently manufactured in Australia and China, it is hoped that as demand grows in other regions of the world production plants will be developed closer to the construction need. The product has been used in affordable housing projects in many regions including Australia, Pacific Island nations and South East Asia.

Conceptos Plasticos based in Colombia also produce a lightweight building material from plastic waste. These blocks are shaped similar to lego pieces so that they slot together, making construction simple. They are also reinforced with a fire retardant to improve their safety and resilience. Anyone can build a house using the bricks and similar to the NevHouse system, the process can normally be completed in just five days. The cost of construction is reported to be 30 per cent cheaper than buildings typically constructed in the region (Cooke, 2016).

Prefabrication

Prefabrication involves manufacturing housing components in bulk in a factory and transporting them to a construction site for assembly. The quality of the building components can be controlled more easily than those manufactured on site, and the fact that components are produced en masse reduces costs. We have already seen some examples of prefabrication such the LGS frame and associated foundations, which are prefabricated and transported to a construction site. We demonstrated how these frames were simple to assemble and the remaining building components can be produced on site from locally sourced materials. Similarly, the NevHouse (NevHouse, 2017) and Conceptos Plasticos (Cooke, 2016) are both examples of prefabricated construction, where all required building components are supplied, and can be easily assembled by non-experts.

There are many ways prefabricating housing components can reduce costs, compared to on-site construction. Labour costs are significantly lower; building materials can be purchased in bulk; and construction processes can be automated. During on-site construction, bad weather can cause significant delays, which also increases costs. At the Druk White Lotus school in Ladakh, India construction can only be performed three months of the year as the weather conditions do not allow construction the rest of the time (Doloi, Green, and Donovan, 2019). Thus, the construction of the school is taking place very gradually over many years. If the pieces of the building could be constructed indoors where they are not affected by weather, then simply transported to the site and assembled construction could occur much more rapidly. During on-site construction, partially built houses will be exposed to bad weather conditions, which could cause damage depending on the building material. For example, wooden frames can be damaged by exposure to moisture during rain. Even without delays prefabrication is still much faster than on-site construction. Both the NevHouse and Conceptos Plasticos reportedly took only five days to construct. For more complex modern style houses, Construction World (2018) reports that it takes half the time to complete a prefabricated house than on-site construction. The quality of prefabricated homes is also more consistent due to the repetitive nature of the manufacturing process.

Prefabrication also has environmental benefits over on-site construction (Construction World, 2018). At construction sites excess materials become waste and are transported to waste processing plants for recycling or disposal. At a prefabrication site, the excess materials can be recycled in-situ, avoiding the pollution associated with transporting waste. This also means it is less likely that excess components will be discarded, as they can simply be used in the construction of other components. The transporting of the preconstructed housing pieces to the construction site usually requires considerably fewer vehicle movements than traditional on-site construction which often involves separate deliveries of all the different building materials constantly over weeks, creating noise and pollution nuisance for nearby residences.

Critics of prefabricated architecture dislike that fact that components are being manufactured for buildings that have not been commissioned. They believe that buildings are site-specific and as such should be built on demand to avoid mass produced building components going unused (Wells, 1986). They also criticise the lack of flexibility in the overall design of prefabricated housing, perceiving that they will look the same inside and out, failing to cater to individual tastes (Quale, 2006). While most appreciate the importance of developing repetitive construction processes to minimise costs, they also see the need to allow a certain amount of flexibility in the design to ensure end-users are satisfied with their house (Dorsey, 1989). Part of this perception stems from the mass-built block towers of former socialist economies such as the USSR. These buildings have come to symbolise the lack of freedom and individuality imposed by oppressive governments.

In contrast, the modern U.S. prefabrication industry is quite well developed, and many high-end homes are built from prefabricated components with plenty of flexibility in the design features. While these products have catered for more expensive houses aimed at middle to high income households, their processes are becoming more cost-effective. A course at Virginia Tech set students a challenge to develop a low-cost prefabricated house for a poor community (Quale, 2006). As part of the project the students were required to consult with members of the community and to incorporate the needs identified through this consultation into the design of the buildings. Despite having to make many alterations to cater to the community's needs the students managed to complete the assignment. After the course, the construction was completed within the limited budget specified, showing that incorporating flexibility into design is possible even for low-cost prefabricated housing. The most important part of the project was consultation with the target community. It was noted that had this happened earlier in the design process, it could have prevented the need to make significant alterations at a later stage.

Repurposing existing buildings

In some regions vernacular buildings have withstood the test of time and remain standing and strong. The desire to preserve historical buildings sees many of these protected from being demolished, however, they are also often no longer used for their original purpose. In Europe for example, where industrialised agriculture means many small-holder farms are no longer functioning, many traditional farm buildings are still standing but serving no purpose other than preservation of history. Fuentes (2010) looked at repurposing these buildings, making them functional for modern purposes. The authors define a six-step approach to repurposing, with the key being to identify the most important vernacular features to ensure these are retained. The scheme also aims to ensure any rebuilding or extension of the buildings is done using traditional building materials and construction techniques.

Labour

Apart from building materials, skilled labour is also essential to develop affordable housing. Many countries experience shortages of skilled labour, for example, the Honourable Jenny Salesa, Minister of Building and Construction recently noted in a presentation at the Pacific Peoples Housing Forum that there is a shortage of 40,000 skilled construction workers in New Zealand (Salesa, 2019). Where skilled labourers are in demand they can charge higher rates, increasing the overall costs of construction. The high demand gives skilled labourers the choice of where to work, and thus they are likely to prefer working in urban areas, where construction jobs are usually higher paid and more consistent. Qualifying for skilled labour often takes years and courses are not likely to be available in all regions, particularly remote rural areas. Not-for-profit organisations such as Habitat for Humanity often bring construction training to remote communities. One of the philosophies of the programme is that if a few members of the local community work together with trained volunteers to build their own homes, they will develop construction skills that they can pass on to other members of the community. Thus, the necessity to import skilled labour from other regions can be avoided and rural communities can become self-sufficient.

Funding land tenure, house construction and policy implications

In this chapter we have looked at many ways land tenure and construction costs can be minimised. While this can go some way to increasing access to housing for low-income families, there will still be many who may need financial assistance to gain access to adequate housing. Government assistance is therefore likely to remain necessary in some form. There are two broad forms this can take: supply side, the government funds the construction of affordable houses and low-income families apply to live in these at heavily subsidised rents (Acolin, 2018). The problem with this approach is that the government continues to own the house, and thus retains responsibility for repairs and maintenance. In the past, governments have failed to accurately estimate the level of these ongoing costs, thus future governments are forced to find room in the budget to not just build more housing but pay for repairs and maintenance to the existing stock. Similarly, as the residents of the property are not responsible for repairs and maintenance they may be less likely to take care of the house. Thus, these houses often end up falling into disrepair.

To gain access to this housing, the government usually has a set of criteria, such as level of income and number of family members, to establish eligibility for the houses. Unfortunately, the number of people fulfilling the criteria has often been underestimated, and the number of houses has fallen well short of the number of people eligible for the scheme. Thus, many people have ended up on waiting lists, taking years to gain a home. Realistically no government has the resources to manage a successful housing scheme in the long term. In fact, it has been argued that the government has no place attempting to do this as it takes away from the housing industry.

The other way for governments to assist low-income families into housing is demand side provision (Acolin, 2018). In this case the government implements policies that allow the creation of affordable houses in the free market. There are many forms of this, which we have already touched on. It could include putting caveats on new housing developments that require a certain percentage to be sold at below market rate to eligible families. It could also include simplifying access to mortgages for families with informal income. There are risks with these sorts of policies, for example, the requirement for houses to be sold at below market value is usually only a short-term policy. After this time, the owners are free to sell it on at market value and thus it will no longer provide an affordable house. Increasing access to mortgages for low-income households is a financial risk. However, if implemented cautiously these schemes can be successful and will be essential to ensure affordable housing goals are met.

Countries are often judged by the quality of life of their citizens, and the quality of their housing will be one of the core components of this. Therefore, it is in the government's interest to implement effective policies to ensure adequate housing is available for people on all levels of incomes in all areas. A major part of this will be finding ways to simplify access to land tenure and ensure the housing stock increases in line with population increase.

References

Acolin, Arthur. (2018). Better location, better housing: Incorporating location into affordable housing loan programs. *Housing Finance International*, 16–24.

Ademiluyi, Israel A., and Raji, Bashiru A. (2008). Public and private developers as agents in urban housing delivery in Sub-Saharan Africa: The situation in Lagos state. *Humanity and Social Sciences Journal*, 3, 143–150.

Australian Government. (2019). Population and households. Australian Government, Australian Institute of Family Studies. Retrieved from https://aifs.gov.au/facts-and-figures/population-and-households.

Averda. (2019). Creating eco-bricks from plastic waste. Averda. Retrieved from https://averda.co.za/news/creating-eco-bricks-plastic-waste/.

Bijlani, H. U. (1982). Rural housing in India. *Habitat International*, 6, 513–525.

ByFusion. (2019). Reshaping the future of plastic. ByFusion Global Inc. Retrieved from www.byfusion.com/.

Central Land Council. (n.d.). The Aboriginal Land Rights Act. Central Land Council. Retrieved from www.clc.org.au/articles/info/the-aboriginal-land-rights-act/.

Chel, Arvind, and Tiwari, G. N. (2009). Thermal performance and embodied energy analysis of a passive house – Case study of vault roof mud-house in India. *Applied Energy*, 86, 1956–1969.

Construction World. (2018). 7 benefits of prefabricated construction. *Construction World*. Retrieved from www.constructionworld.org/7-benefits-prefabricated-construction/.

Cooke, Lacy. (2016). These LEGO-like recycled plastic bricks create sturdy homes for just $5,200. inhabitat. Retrieved from https://inhabitat.com/lego-like-building-blocks-of-recycled-plastic-allow-colombians-to-build-their-own-homes/.

Cool Earth. (2015). The illegal land grab in Papua New Guinea. Cool Earth. Retrieved from www.coolearth.org/2015/09/the-illegal-land-grab-in-papua-new-guinea/.

Dodo, Yakubu Aminu, Ahmad, Mohd Hamdan, Dodo, Mansir, Bashir, Faizah Moham-med, and Shika, Suleiman Aliyu. (2014). Lessons from Sukur vernacular architecture: A building material perspective. *Advanced Materials Research*, 935, 207–210.

Doloi, Hemanta, Green, Ray, and Donovan, Sally. (2019). *Planning, housing and infrastructure for Smart Villages*. Abingdon, U.K.: Routledge.

Dorsey, Robert W. (1989). Integration of architectural and engineering skills. In Oktay Ural and L. David Shen (eds), *Affordable housing: A challenge for civil engineers*. New York: American Society of Civil Engineers.

Dow, Coral, and Gardiner-Garden, John. (1998). Indigenous affairs in Australia, New Zealand, Canada, United States of America, Norway and Sweden. Background Paper 15, Parliament of Australia. Retrieved from www.aph.gov.au/About_Parliament/Parliamentary_Departments/Parliamentary_Library/Publications_Archive/Background_Papers/bp9798/98Bp15.

Dowton, Paul. (2013). "Mud brick." In *Your Home: Australia's Guide to Environmentally Sustainable Homes*, Chris Reardon. Canberra: Department of Climate Change and Energy Efficiency, Australian Government.

Dr. Fixit. (2019). Company profile. Retrieved from www.drfixit.co.in/company-profile.html.

Durand-Lasserve, Alain, and Royston, Lauren. (2002). *Holding their ground: Secure land tenure for the urban poor in developing countries*. London: Routledge.

Fathy, Hasan. (1973). *Architecture for the poor*. Chicago: University of Chicago Press.

Fuentes, José María. (2010). Methodological bases for documenting and reusing vernacular farm architecture. *Journal of Cultural Heritage*, 11. 119–129.

Gabriel, Michelle, Jacobs, Keith, Arthurson, Kathy, Burke, Terry, and Yates, Judith. (2005). National Research Venture 3: Housing affordability for lower income Australians. In *Conceptualising and measuring the housing affordability problem*, Australian Housing and Urban Research Institute (ed.). Melbourne.

Goswami, Monomoy, Varma, Vivek, Dhar, Santosh Mohan, Swargiary, Bitupan, and Boro, Rosey. (2018). A novel approach of constructing ferrocement wall for cost-effective housing. In *1st International Conference on Smart Villages and Rural Development*, Hemanta Doloi, Atul Bora, and Sally Donovan (eds). Guwahati, India: The University of Melbourne.

Hoornweg, D., and Bhada-Tata, P. (2012). What a waste: A global review of solid waste management. In T. W. Bank (ed.), *Urban development knowledge papers*. Washington, D.C.: The World Bank.

International Labour Organization. (1957). C107 – Indigenous and tribal populations convention. United Nations. Retrieved from www.ilo.org/dyn/normlex/en/f?p=NORMLEXPUB:12100:0::NO::P12100_ILO_CODE:C107.

Jeffrey, Craig. (2000). Democratisation without representation? The power and political strategies of a rural elite in north India. *Political Geography*, 19, 1013–1036.

Kumar, J. M. (1962). *Rural housing for India*. Master of Architecture Thesis, University of Manitoba, Canada.

Kumar, Sanjay. (2018). Affordable housing: A case of land ownership in the Andaman and Nicobar Islands. *Housing Finance International*, 43–46.

Lopez, Sarah Lynn. (2010). The remittance house: Architecture of migration in rural Mexico. *Buildings and Landscapes: Journal of the Vernacular Architecture Forum*, 17, 33–52.

Mukhija, Vinit. (2004). The contradictions in enabling private developers of affordable housing: A cautionary case from Ahmedabad, India. *Urban Studies*, 41, 2231–2244.

Nath, Arun Jyoti, Das, Gitasree, and Das, Ashesh Kumar. (2009). Above ground standing biomass and carbon storage in village bamboos in North East India. *Biomass and Bioenergy*, 33. 1188–1196.

NevHouse. (2017). NevHouse: Housing humanity. Retrieved from www.nevhouse.com/.

Quale, J. (2006). Ecological, modular and affordable housing. *WIT Transactions on the Built Environment*, 86, 53–62.

Ramesh, T., Prakash, Ravi, and Shukla, K. K. (2012). Life cycle energy analysis of a residential building with different envelopes and climates in Indian context. *Applied Energy*, 89, 193–202.

Roach, Mary. (1996). The bamboo solution. Retrieved from http://discovermagazine. com/1996/jun/thebamboosolutio784.

Salesa, J. (2019). Opportunities and challenges in Pacific housing. Pacific Peoples Housing Forum, 17 May 2019, Auckland, New Zealand.

Sazinski, Richard J. (1989). Lower cost structural techniques for housing. In Oktay Ural and L. David Shen (eds), *Affordable housing: A challenge for civil engineers*. New York: American Society of Civil Engineers.

Statista. (2019a). Average number of people living in households in China from 1990 to 2017. Statista. Retrieved from www.statista.com/statistics/278697/average-si ze-of-households-in-china/.

Statista. (2019b). Average number of people per household in the United States from 1960 to 2018. Statista. Retrieved from www.statista.com/statistics/183648/avera ge-size-of-households-in-the-us/.

Tanabe, Akio. (2007). Toward vernacular democracy: Moral society and post-post-colonial transformation in rural Orissa, India. *American Ethnologist*, 34, 558–574.

United Nations, Department of Economic and Social Affairs, Population Division. (2017). Household size and composition around the world 2017 – Data booklet. United Nations, ST/ESA/SEr.A/405.

Walker, Kathy Le Mons. (2008). Neoliberalism on the ground in rural India: Predatory growth, agrarian crisis, internal colonization, and the intensifictaion of class struggle. *The Journal of Peasant Studies*, 35, 557–620.

WasteAid UK. (2019). How to turn mixed plastic waste and bottles into ecobricks. WasteAid UK. Retrieved from https://wasteaid.org/toolkit/how-to-turn-mixed-pla stic-waste-and-bottles-into-ecobricks/.

Wells, Jill. (1986). *The construction industry in developing countries: Alternative strategies for development*. Beckenham, U.K.: Croom Helm.

World Habitat. (2017). Bringing light and air to homes in informal settlements. Retrieved from www.world-habitat.org/world-habitat-awards/winners-and-finalists/ bringing-light-and-air-to-homes-in-informal-settlements/.

7 Global practices in rural development

This chapter offers an overview of best practices in rural development that are built around access to housing in rural areas. While considerable literature on rural housing and rural development exists in developed countries, such topics have not been fully explored in developing countries. In this chapter, the focus is on practices in India that may be relevant in promoting rural development in other developing countries.

In 2015 the Government of India launched an affordable housing scheme entitled Pradhan Mantri Awaas Yojana, more commonly known by its acronym PMAY. The scheme comprised separate components for urban (PMAYU) and rural (PMAYR or PMAYG). It is an extension of the Indira Awaas Yojana scheme initially launched in 1985. The scheme aims to build 20 million houses by the end of 2022. The scheme will of course aid in the attainment of the United Nations Sustainable Development Goal 11: Sustainable Cities and Communities, whose first target is "By 2030 ensure access for all to adequate, safe and affordable housing" (United Nations, 2019). However, the scheme also incorporates various other development goals including:

Goal 6: Clean Water and Sanitation – houses built under the PMAYG scheme are required to include a sanitary toilet (one that prevents contact with faecal matter) and access to drinking water;
Goal 7: Affordable and Clean Energy – houses built under PMAYG are required to have a gas and electricity connection;
Goal 5: Gender Equality – ownership of the houses is granted to the woman or jointly between husband and wife.

Apart from these the scheme will also aid in reducing poverty (Goal 1), improving health and well-being (Goal 3) and reducing inequalities (Goal 10).

Therefore the scheme intends to reach beyond the basic provision of shelter to bring genuine improvement to the quality of life of rural residents in India.

Unfortunately, the scheme has been criticised for the generic design of the houses and the use of expensive imported building materials such as brick and concrete. The failure of the scheme to incorporate vernacular designs and locally available renewable building materials has raised concerns of both climatic responsiveness and cultural suitability of the housing. In response to this we conducted a series of interviews with residents in rural areas of the state of Assam, some in houses built under PMAYG and some in self-built houses. A summary of these survey results is shown in Table 7.1.

Government supported PMAYG houses, India

The basis of the PMAYG housing was the compendium of rural housing typologies known has Pahel (UNDP, 2017). Based on a study conducted across 18 states in India, the Ministry of Rural Development, India developed a range of housing layouts and designs incorporating 130 different climatic zones across the states. The idea was to integrate local conditions and resources in order to develop houses that were fit-for-purpose from the community perspective.

We interviewed people from nine households that had been built under the PMAYG scheme in 2017 or 2018. Out of these five had been modified at least twice since the original construction of the house, and one had been modified four times, the largest number of modifications. All of the people had been granted 130,000 rupees under the scheme, but all but one had invested some of their own income into the construction as well. These personal investments ranged from 18,000 to 200,000 rupees. All of the residents said they were happy with their house although two stated that the construction was incomplete and required some more plastering. We also asked the residents whether, if they could afford to build their own house, they would build in the same style as the house provided by the scheme. Only four said yes, with five saying no they would build a different style of house. However, all nine said they would expect government support for any building they undertook.

A summary of these interviews along with a photograph of the respective houses (Figures 7.1–7.9) are presented in the following case studies.

Case study P.1: PMAYG housing scheme beneficiary, Assam, India (Resident P.1)

Resident P.1 is a beneficiary of Pradhan Mantri Adarsh Gram Yojana (PMAYG) which is a rural development programme launched by the Central government in India in the financial year 2009–10 for the development of villages having a population with 50 per cent or more citizens with Schedule Castes status. The size of the house is approximately 54 m2 and was built in 2017 at a cost of Rs. 130,000 (AUD 2600) received from PMAYG. In addition to the government fund, approximately Rs.

50,000 (AUD 1000) was spent from his own resources to complete the house. The piece of land was inherited from relatives and is approximately 2 Kathas (540 m2). The head of household's primary income generating activity is tailoring and they earn about Rs. 300 per day (monthly Rs. 7,200 ~ AUD 144). The house generally meets the needs of the family, though some work such as rendering is yet to be done. Although they appreciated the government funded house, they would have preferred a different house design if they had the option.

Figure 7.1 PMAYG house of Resident P.1

Case study P.2: PMAYG housing scheme beneficiary, Assam, India (Resident P.2)

Resident P.2 is a beneficiary of the PMAYG scheme where he was awarded a house of approximately 35 m2, built in 2017 at a cost of Rs. 130,000 (AUD 2600). An additional sum of Rs. 30,000 (AUD 600) was invested in the house from his own income. The fund was released in three instalments. The piece of land is approximately 1.5 Kathas (405 m2) inherited from the family. The primary house owner is a Rickshaw (three wheeled push trailer) driver and earns approximately Rs. 300 per day (monthly Rs. 7,200 ~ AUD 144). The house is generally meeting the needs of the family, but if alternative funding was available, the owner would have preferred to build a different style of house.

Figure 7.2 PMAYG house of Resident P.2

Case study P.3: PMAYG housing scheme beneficiary, Assam, India (Resident P.3)

Resident P.3 is a beneficiary of PMAYG scheme where she was awarded a house of about 16 m2, built in 2018 at a cost of Rs. 130,000 (AUD 2600). No additional sum was invested in the house from the owner, but the house is considered incomplete. Both the render and windows need some work. The piece of land is approximately 1.5 Kathas (405 m2) inherited from the family. The primary house owner is an unskilled day labourer, earning approximately Rs. 50 per day (monthly Rs. 3,500 ~ AUD 70). The house is generally meeting the needs of the family, and if the government funding was not available, the owner would not been able to afford this house in her lifetime.

Figure 7.3 PMAYG house of Resident P.3

Case study P.4: PMAYG housing scheme beneficiary, Assam, India (Resident P.4)

Resident P.4 is a beneficiary of PMAYG scheme where he was awarded a house of about 35 m2, built in 2017 at a cost of Rs. 130,000 (AUD 2600). An additional sum of Rs. 30,000 (AUD 600) was invested in the house from his own income. The fund was released in three instalments. The piece of land is approximately 1.5 Kathas (405 m2) inherited from the family. The primary house owner is a farmer and earns a basic income for supporting his family. The income is dependent on the harvesting season and yields of the day. The house is generally meeting the needs of the family. If government funding was not available, the owner would not have been able to afford such a house on his own, so he is happy with this house.

Figure 7.4 PMAYG house of Resident P.4

Case study P.5: PMAYG housing scheme beneficiary, Assam, India (Resident P.5)

Resident P.5 is a beneficiary of the PMAYG scheme where he was awarded a house of about 14 m2, built in 2018 at a cost of Rs. 130,000 (AUD 2600). This new house replaced his old bamboo house of approximately 11 m2. An additional sum of Rs. 33,000 (AUD 660) was invested in the house from his own income. The fund was released in three instalments. The piece of land is inherited from the family. The primary house owner is an unskilled day labourer and earns a basic income of about $250 per month for supporting his family. The house is generally meeting the needs of the family and if the government funding was not available, the owner would not have been able to build this house. However, given the choice they would have preferred a different style of house.

Figure 7.5 PMAYG house of Resident P.5

Case study P.6: PMAYG housing scheme beneficiary, Assam, India (Resident P.6)

Resident P.6 is also a beneficiary of the PMAYG scheme. He upgraded his bamboo house of 20 m2 with a brick and mortar house of roughly the same size. The new house was built in 2017 at a cost of Rs. 130,000 (AUD 2600). Since 2017, the house has been extended twice with an additional personal cost of Rs. 200,000 (AUD 4000) to make it suitable for the family. The fund was released in three instalments. The piece of land is inherited from the family. The primary house owner is an unskilled day labourer and earns a basic income of about $200 per month for supporting his family. The house after modification is generally meeting the needs of the family and if the government funding was not available, the owner would not have been able to build this house in the first place.

Figure 7.6 PMAYG house of Resident P.6

Case study P.7: PMAYG housing scheme beneficiary, Assam, India (Resident P.7)

Another beneficiary of the PMAYG is Resident P.7 who was awarded a house of approximately 52 m^2 replacing an old bamboo house of approximately 42 m^2. This new house was built in 2017 at a cost of Rs. 130,000 (AUD 2600). An additional sum of Rs. 18,000 (AUD 360) was

Figure 7.7 PMAYG house of Resident P.7

invested in the house from his own income. The fund was released in three instalments. The piece of land is inherited from the family. The primary house owner is a farm worker and does not have any disposable regular income. The house is generally meeting the needs of the family, and if the government funding was not available, the owner would not have been able to build this house.

Case study P.8: PMAYG housing scheme beneficiary, Assam, India (Resident P.8)

Resident P.8 is a beneficiary of the PMAYG scheme. She upgraded her 15 m2 bamboo house with a brick and mortar house of almost double the size at 25 m2. The new house was built in 2017 at a cost of Rs. 130,000 (AUD 2600). An additional personal cost of Rs. 60,000 (AUD 1,200) was spent to make it suitable for the family. The fund was released in three instalments. The piece of land is inherited from the family. The primary house owner is a farm labourer and does not have any regular income. The family is dependent on the farm produce. The house after modification is generally meeting the needs of the family. Even without government intervention, the owner still would have built such a house for her family.

Figure 7.8 PMAYG house of Resident P.8

Case study P.9: PMAYG housing scheme beneficiary, Assam, India (Resident P.9)

Resident P.9 is a beneficiary of the PMAYG scheme where he was awarded a house of approximately 25 m2. The house was built in 2017 at a cost of Rs. 130,000 (AUD 2600). This new house replaced his old bamboo house of size approximately 16 m2. An additional sum of Rs. 20,000 (AUD 400) was invested in the house from his own income. The fund was released in three instalments. The piece of land is inherited from the family. The primary house owner being farm worker does not have regular income but depends on seasonal crop yields. The house is generally meeting the needs of the family, but if the government funding was not available, the owner would have built a different style of house.

Figure 7.9 PMAYG house of Resident P.9

Self-made houses, India

We also interviewed a further six people in the same region who were living in a house that was inherited from a relative and one who had purchased an existing house. Four of the people living in these houses felt that they were too small and hoped or were already planning to extend their home. The main reason for extension was to add more bedrooms.

We also interviewed three people who had self-funded the construction of their houses. The construction costs of one of these houses was 20,000 rupees, much cheaper than the allotted 130,000 from the PMAYG scheme. The owners were hoping to extend the house in the future. The other two cost 350,000 rupees, the owners of both of which were happy with the house and did not feel it needed any extensions or improvements. The jobs of these householders had greater incomes than those of the other groups, thus they would not qualify as poor, but they are included here for comparative purposes. All of the self-built houses had a detached toilet with septic tank, access to water either from wells or the public health supply on tap from a central location. However, none of them had electricity access and relied on LPG bottles for cooking.

These interview transcripts along with the respective houses (Figures 7.10–7.19) are presented as case studies in the following sections.

Case study S.1: self-made home, Assam, India (Resident S.1)

Resident S.1, from a village called Manahkuchi, Hajo, Kampur, built his first "Kacha Ghar" house in 1979. The house comprises three bedrooms, one kitchen and one front veranda. The piece of land is approximately 2 Kathas (540 m2) inherited from the family. Being a day labourer for low-skilled household work, the primary resident earns about Rs. 2100 per week (AUD 42) and built the house about for about Rs. 20,000 (AUD 400). While the house is currently serving the purpose of the family well, there is a plan to extend one or two bedrooms in the future.

Figure 7.10 Self-made house of Resident S.1

Case study S.2: self-made home, Assam, India (Resident S.2)

Resident S.2, from a village called Nakuchi, Hajo, in Kampur district, built his first "Reinforced Concrete Cement" (RCC) house in 2012. The house was built over seven months, comprises 11 bedrooms, one garage, one store room, two office rooms, one kitchen, one drawing room and four balconies. The piece of land is approximately 4 bighas (21,6000 m2) received by inheritance. Being a business man, the primary resident earns about Rs. 600,000 per month (AUD 12,000). The cost of the house is approximately Rs. 5,000,000 (AUD 100,000).

Figure 7.11 Self-made house of Resident S.2

Case study S.3: self-made home, Assam, India (Resident S.3)

Resident S.3, from the village of Duliagaon, Garamur in Jorhat district, built his first "Assam Type" house in 1998. The house was built within a

Figure 7.12 Self-made house of Resident S.3

year and comprises four bedrooms, one guest room, one kitchen-dining room and two front and rear verandas. The piece of land is approximately four kathas (1,080 m2) and was bought on his own in 1972 at a cost of Rs. 20,000 (AUD 400). Being the primary owner, he constructed the house for approximately Rs. 2,10,000 (AUD 4200). He is a government employee with an average salary of Rs. 21,000 (AUD 420) per month. The whole family is extremely happy living in this house and there is no any need for alteration or further extension in any part of the house.

Case study S.4: self-made home, Assam, India (Resident S.4)

Resident S.4 is from the village of Duliagaon, Garamur in Jorhat district and lives in an "Assam Type half wall" house that was built by his grandfather in 1950. The house was later modified in 2000 with a new tin roof for about Rs. 40,000 (AUD 800). The house comprises four bedrooms, two guest rooms, one prayer room, two verandas and one kitchen-dining room. The piece of land is approximately one bigha (5400 m2) and was inherited from their forefather. The primary resident is a rice farmer and earns some basic income to support the family, but the exact amount of income is not known. The whole family is happy living in this house but there is a plan to add one or two bedrooms in the near future.

Figure 7.13 Self-made house of Resident S.4

Case study S.5: self-made home, Assam, India (Resident S.5)

Resident S.5, from a village called Kumar Gaon, Garamur in Jorhat district, lives in an "Assam Type half wall" house that was built by his father in 1961. The original house was a "Kucha Ghar" but was later modified in 2005 to a "Assam Type half wall" house. The modification was done in two stages. In the first stage, a cost of approximately Rs. 150,000 (AUD 3,000) was invested. Then about a year later, an additional Rs. 250,000 (AUD 5,000) was invested to complete the renovation. The house comprises three bedrooms, one prayer room, two verandas and one kitchen-dining room. The piece of land is approximately one bigha and 1.5 kathas (5,805 m2) and was bought by his father in 1960 for an amount of Rs. 800 (AUD 16). Being a small private businessman, the primary resident earns about Rs. 10,000 (AUD 200) per month. The whole family is happy living in this house but there is a plan to extend it by adding one or two bedrooms in the near future.

Figure 7.14 Self-made house of Resident S.5

Case study S.6: self-made home, Assam, India (Resident S.6)

Resident S.6, from a village called Manahkuchi Hajo in Kamrup district, lives in a "Kucha Ghar" house that was built by in 1979. The house comprises three bedrooms, one kitchen and one veranda. The piece of land is approximately four kathas (1080 m2) and was inherited from his father. Being a small rice farmer, the primary resident earns a small amount of money by selling rice. The cost of construction of the house was Rs. 30,000 (AUD 600). The whole family is happy living in this house but there is a plan to extend the house by adding one or two bedrooms in the near future.

Figure 7.15 Self-made house of Resident S.6

Case study S.7: self-made home, Assam, India (Resident S.7)

Resident S.7, from a village called Namjamguri in Jorhat district, lives in a "Kucha Ghar" house that was built in 2005. The house was built at a cost

Figure 7.16 Self-made house of Resident S.7

of approximately Rs. 20,000 (AUD 400) and comprises three bedrooms, one drawing room and one kitchen. The piece of land is approximate two bighas (10,800 m2) and was inherited from an ancestor. Being a day labourer, the primary resident earns about Rs. 8400 (AUD 168) per month. The whole family is happy living in this house but there is a plan to extend the house in the future.

Case study S.8: self-made home, Assam, India (Resident S.8)

Resident S.8, from a village called Duliagaon, Garamur in Jorhat district, lives in an "Assam Type full wall" house that was built by his father in 1985. The original house was modified in 2002 by adding two bedrooms at an approximate cost of Rs. 240,000 (AUD 4800). In the first stage, the total cost was approximately Rs. 60,000 (AUD 1,200). The piece of land is approximately six kathas (1,620 m2) and was bought in 1982 for an amount of Rs. 3,300 (AUD 606). Being a government employee, the primary resident used to earn about Rs. 25,000 (AUD 500) per month but he is retired now. The whole family is happy living in this house.

Figure 7.17 Self-made house of Resident S.8

Case study S.9: self-made home, Assam, India (Resident S.9)

Resident S.9, from a village called Garamur Kumargaon in Jorhat district, lives in an "Assam Type full brick wall" house built in 2011. The total cost of construction was Rs. 400,000 (AUD 8000). The house comprises four bedrooms, one kitchen, one drawing room and one verandah. The piece of land is approximately two kathas (540 m2) and was bought in 2006 at a cost of Rs. 150,000 (AUD 3,000). Being a government employee, the primary resident earns about Rs. 30,000 (AUD 600) per month. The whole family is happy living in this house and there is no any plan or need for extending the house any more.

Figure 7.18 Self-made house of Resident S.9

Case study S.10: self-made home, Assam, India (Resident S.10)

Resident S.10, from a village called Garamur Kumargaon in Jorhat district, lives in an "Assam Type" house built in 2014 over a period of seven months. The cost of construction was Rs. 350,000 (AUD 7000). The house was extended in 2016 by investing another Rs. 150,000 (AUD 3000). The house comprises two bedrooms, one guest room, one drawing cum dining room, one kitchen, two verandahs and one prayer room. The piece of land is approximately five kathas (1350 m2) and was bought in 1960 at a cost of Rs. 16,000 (AUD 320). Being a small tea grower, the primary resident earns about Rs. 60,000 (AUD 1,200) per month. The whole family is happy living in this house and there is no any plan or need for extending the house any more.

Figure 7.19 Self-made house of Resident S.10

Summary analysis of all the case studies presented above is shown in Table 7.1.

The self-built houses fell into three categories of building type – one was a modern house predominantly made of concrete, three were classified as Kacha Ghar, which refers to houses made from locally available renewable materials such as mud, dry grass and wood. These houses are cheap and quick to construct and are often considered poor quality. However, Kacha houses often display superior climatic responsiveness and although they are less likely to withstand strong winds and heavy rainfall, they are less dangerous to residents if they do break in a storm and they are relatively cheap and easy to rebuild. We will talk in more detail about resilience in building in Chapter 11.

The others were classified as Assam Type houses which were developed by the British who occupied the state of Assam for 120 years from 1826 (Khandekhar, 2017). Prior to this, the buildings had been mainly Kacha style unless they had been built for kings in which case they were more complex, cumbersome buildings that did not respond to the local climatic conditions. Figure 7.20 shows a typical Kacha house in a rural Assamese village.

The British are reported to have developed the Assam Type style in response to the local environment, climatic conditions and available building materials. Although considered vernacular architecture the Assam Type house comprises many features that comply with the contemporary preference for sustainability in buildings.

Table 7.1 Results of survey questions from Assam, India

PMAYG Interview results

Resident	Area	Year house built	Original house size	New house size	Amount	Modifications?	Land tenure	Any self-funding?	Occupation	Income (Rupees per day)	Does house need work?	If so what?	Would you build same again?	Do you expect government funding for future building?
P.1	Garmur, Near Green Park	2017	6.7	8.2	130000	4	Self-owned	50,000	tailor	300	Yes	Plastering	Yes	Yes
P.2	Baligaon, Jorhat	2017	3.7	4.3	130000	0	Self-owned	30000	rickshaw driver	275	No		No	Yes
P.3	Baligaon, Jorhat	2018	3.7	3.7	130000	2	Self-owned	0	Labour	50	Yes	plastering, windows	Yes	Yes
P.4	Bakigaon, Jorhat	2018	4	3.7	130000	3	Self-owned	30000	Farmer	Unpredictable	No		Yes	Yes
P.5	Bakigaon, Jorhat	2018	4	3	130000	3	Self-owned	33000	labour	350	No		No	Yes
P.6	Borkolia Mising Gaon	2017	3.7	5.5	130000	2	Self-owned	200000	Labour	300	No		No	Yes
P.7	Bakigaon, Jorhat	2017	4	9.7	130000	0	Self-owned	18000	Farmer	Unpredictable	No		No	Yes

Resident	Area												
P.8	Mising Gaon, Majuli	2017	4	3.7	130000	0	Self-owned	60000	Farmer	Unpredictable	No	Yes	Yes	Yes
P.9	Bakigaon, Jorhat	2017	4	4.3	130000	0	Self-owned	20000	Farmer	Unpredictable	No	No	Yes

Self-built house survey responses

Resident	Area	Year house built	House size	How gained?	Value (Rupees)	Modifications?	Land tenure	Occupation	Income (Rupees per day)	Does house need work?	If so what?	House style
S.1	Manahkuchi, Hajo, Kampur	1979	3.6	Inherited	20000	0	Self-owned	Labour	300	Yes	Extension adding more bedrooms	Kacha Gar
S.2	Nakuchi, Hajo, Kampur	2012	14.5	Inherited	5000000	0	Self-owned	Business	21,000			RCC
S.3	Duliagon, Garamur, Jorhat	1998	6.3	Purchased 1972	210000	0	Self-owned	Government	750	No		Assam Type
S.4	Duliagon, Garamur, Jorhat	1950	7.2	Inherited	-	1	Self-owned	Farmer	Unpredictable	Yes	Extension adding more bedrooms	Assam Type half-wall

S.5	Kumar Gaon, Garamur, Jorhat	1961	5.4	Inherited	-	2	Self-owned	Business	357	Yes	Extension adding more bedrooms	Assam Type half-wall
S.6	Manhkuchi Hajo, Kamrup	1979	4	Inherited	30000	0	Self-owned	Farmer	Unpredictable	Yes	Extension adding more bedrooms	Kacha Gar
S.7	Namjamguri, Jorhat	2005	4.5	Self-funded	20000	0	Self-owned	Labour	300	Yes	Extension	Kacha Gar
S.8	Duliagon, Garamur, Jorhat	1985	3.5	Inherited	60000	1	Self-owned	Retired former government employee	893	No	Extension	Assam Type full-wall
S.9	Garamur Kumargaon, Jorhat	2011	6	Self-funded	350000	0	Self-owned	Government	600	No		Assam Type full-wall
S.10	Garamur Kumargaon, Jorhat	2014	4.5	Self-funded	350000	1	Self-owned	Tea grower	571	No		Assam Type full brick wall

Figure 7.20 Typical Assamese Kacha house

Firstly, they are usually built on a raised plinth, making the floor of the house approximately 60–70 cm above ground level. This serves multiple functions including reducing the amount of dust and smoke entering the house from the roads outside; resilience against flooding which is prevalent in region; and limiting access to wildlife, especially snakes that could pose a risk to the house's occupants. The plinth is generally constructed from more modern building materials such as brick and concrete. However, the actual house structure is built using locally available, renewable building materials. The house is made with a timber frame, usually constructed from Sal wood, a timber with similar properties to teak commonly found in the region (Kaushik and Babu, 2009). The walls consist of a bamboo frame filled in with ekra (also known as ikra or ikara) a type of grass that grows easily and abundantly. While in the Assam region ekra is seen as a useful material, in many regions it is considered a weed due to how easily it thrives (Wikipedia, 2019). The ekra is plastered over with a mortar made from mud in three layers. The final layer is mud mixed with cow dung (Khandekhar, 2017). The idea of using locally sourced, renewable materials where possible, only incorporating more expensive, imported modern materials such as concrete and brick ties in with the principles of sustainability we discussed in earlier chapters as it keeps costs to a minimum and reduces the embedded greenhouse gas emissions in the building. Figure 7.21 shows a typical vernacular Assam Type house in a rural Assamese village.

Figure 7.21 Vernacular Assam Type house

Perhaps the defining feature of the Assam Type house is the roof, which is always built at the same angle of inclination. Originally the houses had a thatch roof, also made from ekra, although in more recent times these have largely been replaced by corrugated iron sheeting. The use of thatch started to go out of fashion due to the advent of electricity, which could easily spark a fire with a thatch roof due to the frequency of seismic activity in the region. The inclination of the roof is important for providing protection against the heavy rainfall experienced during the annual monsoon season. Thatch roofs are built with a steeper slope than metal sheets to account for the lower tolerance for heavy rainfall of the thatch. The roofs are also built with wide eaves to protect the walls from heavy rainfall. The mud-dung plaster becomes brittle from the heat in the summer, making it easy to wash away during the rainy season. Therefore keeping the rain off the walls is important for the longevity of the building. The houses often require replastering annually, after the monsoon season due to damage to the plaster from the rain. The design of the building includes openings to provide nat-ural ventilation, helping to maintain comfortable interior temperatures during the hot summer season.

Being a vernacular design the Assam Type house has climatic responsive features capable of maintaining comfortable indoor temperatures without the need for heating or air conditioning, although this situation may be challenged in the future as temperatures become more extreme.

The main problem with the houses provided by the government scheme is the high-level management of the building designs. Allowing the building design at a more local scale could allow areas to incorporate the widely used and well understood vernacular architectural features that are often found to have superior performance regarding the local climatic and environmental conditions, while also retaining cultural significance. The preference for the use of the locally sourced renewable materials would also lead to significant cost reductions in the building construction, while the PMAYG scheme houses all cost at least 130,000 rupees to construct, one of the Assam Type houses reportedly cost just 20,000 rupees. The fact that many of these vernacular styles are developed from members of the local communities means funding from the scheme could be used to increase construction jobs for locals, and save the cost of having to bring in builders from further away as local builders may not be familiar with the use of the modern building materials and techniques used in the schemes house designs. This renewed preference for vernacular styles has been seen in other places such as the bamboo housing project in Costa Rica, mentioned briefly in Chapter 6. In 1987 700 affordable houses were built out of bamboo. The cost of each house was approximately US$4500 and was so successful that it was continued at a rate of 1000 houses per year (Roach, 1996).

Best practice for developing affordable housing and policy interventions

While development of affordable housing is a global phenomenon, there are significant variations in concept, theory and practices among the community, academia, practitioners and policy makers. As community is the beneficiary of affordable housing schemes, the affordability equation cannot be solved without placing community at the core of policy-making. The PMAYG housing scheme case studies presented in the chapter highlight that while subsidised community housing programmes are considered useful from the government perspective, the truth is that such programmes have contributed very little or none in upgrading the lifestyle of rural people. It was found that the self-funded houses were equally comfortable for rural lifestyles as those provided by the government scheme. Thus, there is capacity within the community to construct their own housing.

As found in the survey of the self-made homeowners, there are clear aspirations for creating purpose-built houses by individual owners within the close-knit communities to which they belong. The prevalent concept in developed nations of purchasing an affordable house and land package is not necessarily applicable to developing regions. As evidenced in the case studies of the self-made houses, most of the time the house-owner is able to access a piece of land through either inheritance or purchase, then proceeds to build the house in stages over a long period of time. This progressive expansion of houses usually corresponds to growth of the family with no consideration of construction costs or impact on property value. This is clearly a significant point of distinction in the perception of property values, value-for-money or growth potentials among the developed communities from the thinking of a basic shelter and leading a simple lifestyle among the developing rural communities.

In addressing the affordability crisis in housing generally, the responsibility of the government is still very high. However, the traditional intervention process of top-down policies for funding, implementing and delivering of housing programmes for communities is not necessarily effective. Public housing programmes are usually very expensive, and the funding source is largely from tax-payer's contributions. Public policies should be such that the scheme not only delivers cost effective houses that fit in the local conditions but also make the beneficiary community realise the benefits of public goods that result in substantially upgrading living conditions. While the public benefits may come for free, the indirect benefits must be realised and visually evident in empowering the communities and supporting them to lead sustainable lifestyles.

In all the case studies considered in this chapter, both PMAYG and self-built, the residents already had access to a piece of land to build on. Schemes that target the landless will also be an important consideration. One way to achieve this is to decouple land provision from housing provision, cost of construction from growth potential and derivation of value for money with short-term investing planning that will create long-term economic benefits. Public policies should recognise the need to make land available for the community so that housing can be developed without land being a dependent entity.

The most appropriate role for the government to take in the provision of affordable housing may not be in the front line of activities. Most governments lack both the expertise and the capacity to deliver high quality building construction that complies with the appropriate building codes of the region. A more appropriate role for the government may be to get involved with regulatory and enforcing responsibilities for private companies, encouraging them to operate in a fair and reasonable competitive environment.

References

Kaushik, H., and Babu, K.S.R. (2009). *Housing report: Assam-type house.* World Housing Encyclopedia, Report 154, Earthquake Engineering Research Institute and International Association for Earthquake Engineering.

Khandekhar, Y.S., Rahate, O.P., Gawande, A.B., Sirsilla, K.A., and Govindani, S.M. (2017). Vernacular architecture in India. *International Research Journal of Engineering and Technology*, 4(5), 2747–2751.

Roach, Mary. (1996). The bamboo solution. Retrieved from http://discovermagazine.com/1996/jun/thebamboosolutio784.

United Nations. (2019). *Sustainable Development Goals.* New York: United Nations. Retrieved from https://sustainabledevelopment.un.org/?menu=1300.

United Nations Development Programme (UNDP). (2017). *Pahal: A compendium of rural housing typologies.* UNDP in collaboration with Ministry of Rural Development, India and Indian Institute of Technology, Delhi.

Wikipedia. (2019). Pradhan Mantri Awas Yojana. Retrieved from https://en.wikipedia.org/wiki/Pradhan_Mantri_Awas_Yojana.

8 Vulnerability in rural communities

The United Nations' definition of a disaster is "an event that critically disturbs the functioning of a society, causing human, material and environmental damage and making it difficult for societies to resume normal conditions without intervention" (Haigh et al., 2016, 570). Disasters affect people in both developed and developing regions; however, lifestyles of many developing rural communities are already synonymous with hardship and poverty, making it even more difficult for them to resume normal conditions in the wake of a disaster. Houses are often more vulnerable to damage due to poor building quality and low-quality lightweight building materials. While they may have been built well initially, due to lack of time and resources they are rarely maintained, making them less resilient and more vulnerable over time. It is also important to remember that it is not just the building itself that is damaged, but all the possessions contained within can also be lost. This can include all the money of households that do not have access to a banking system and keep all their income in cash. It can mean loss of a long-term supply of food for some families who may have been storing non-perishables. For households where the residents work from home it can mean the loss of their business premises and associated assets.

Disaster is sometimes caused by natural occurrences including seismic activity which causes earthquakes. In the second section of this chapter we will describe the impact of some of the worst earthquakes in recent times on housing in rural communities. Seismic activity can also affect the movement of ocean water creating tsunamis, which can spell disaster in coastal areas. In section three we look at the devastation caused by some of the most famous tsunamis to have occurred.

Severe storms, also known as hurricanes, typhoons or cyclones, can also have a devastating impact on rural housing. While some regions have always experienced these extreme storms, climatologists have noted that in recent times, due to increased ocean temperatures, the frequency of these storms is reduced, but the severity increased. This has been linked to the accumulation of greenhouse gases in the atmosphere. Disturbingly, their projections show the severity of these storms is likely to continue to increase (Lefale,

Diamond, and Anderson, 2018). Section four will look at the impacts of some of these severe storms.

The surrounding landscape of houses can also be severely damaged during a disaster, which is particularly bad news for agricultural households. The loss of crops and livestock can spell loss of both their own supply of food and their income. Vulnerability of agricultural households to disaster is exacerbated by the modern preference for monoculture, where a single crop is grown in the same soil season after season. Rather than working in harmony with nature, monoculture agriculture depletes soils of nutrients, increasing the need for synthetic and mineral fertilisers. Pests tend to be attracted to specific types of crops, so by growing the same crop over and over again the pests that attack that crop will thrive, increasing the need for pesticides and the risk of pesticide resilience.

The deterioration of soils impacts their water retention properties, making them vulnerable to changes in rainfall patterns. Where rainfall is unexpectedly low, droughts occur, resulting in unexpectedly low harvests and increased risk of wildfires. Where rainfall is unexpectedly heavy, soils are eroded, and less water is absorbed into soils increasing the risk of flooding to nearby buildings. In the fifth, sixth and seventh sections of this chapter we will look at examples of these three natural disasters. Figure 8.1 depicts the some of the common natural disasters that can contribute to community vulnerability resulting from sub-standard quality of housing construction. A good quality house fitted with relevant safety features could potentially reduce the impact and vulnerability and thus such a consideration is important in housing design and construction.

While some of these natural disasters have been linked to anthropogenic activity, especially the greenhouse gases emitted by burning fossil fuels,

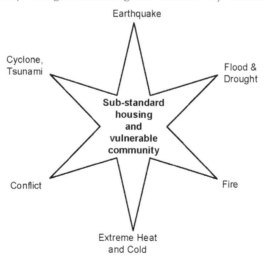

Figure 8.1 Vulnerability associated with sub-standard housing

another more direct form of anthropogenic generated disaster is conflict. Conflicts can have a similar or even worse impact on rural communities than disasters caused by extreme weather events and seismic activity. The eighth section of this chapter will look at specific examples of the way some rural communities have been impacted by conflict.

After a disaster, many houses are severely damaged and may need significant repairs or rebuilding from scratch. In rural communities, where many residents are self-employed, particularly in agriculture, most people will not only lose their home but their source of income. Thus, they are stuck as they have nowhere to live and no way to generate an income to pay for a new home. Fortunately, many third sector organisations are available who are dedicated to house construction and they are always willing to step in with volunteers to assist in the rebuilding process. In this chapter, as we describe the impacts of some of the worst disasters to have occurred in recent times, we will also describe some of the rebuilding efforts that assisted these communities. Then we will conclude the chapter by discussing some of the things that help and some that hinder post-disaster rebuilding efforts.

Earthquakes

The outer layers of the Earth, known as the mantle and crust, are not a continuous form, but rather are made up of fragments known as tectonic plates. These jagged-edged plates are in constant motion, with different plates moving in different directions. Every now and again sections on the edges of these plates get stuck together while the rest of the plate keeps moving (Wald, 2019). Pressure builds at this stuck point until finally the edges become unstuck and the pressure is released, sending a ripple of energy through the Earth's crust in all directions. As a result, the ground moves in what is known as an earthquake. Despite much research into the motion of tectonic plates there is currently no way to predict when earthquakes are likely to occur; experts have only been able to identify locations where they are most likely to occur. These are locations in close proximity to where the edges of tectonic plates meet and are known as fault lines. In this section we describe the impacts of some of the worst earthquakes in recent times.

Turkey

An earthquake of magnitude 7.1 shook an area of approximately 3000 m^2 around the Gediz River, Turkey on 28 March 1970 (Ambraseys and Tchalenko, 1970). Over 250 villages were affected with an estimated 3500 houses destroyed and another 18,000 experiencing varying levels of damage. As a result, many people were left homeless – estimates range from 50,000 (Cromley, 2008) to 80,000 (Ambraseys and Tchalenko, 1970). The predominant religion in the area was Islam, and agriculture was the main form of occupation. Prior to the earthquake the vernacular architecture reflected this. The buildings were mainly two storeys, the ground floor for housing livestock and the upper floor for families. In the Islamic tradition, the houses had high levels of privacy and consisted of individual rooms built around a central courtyard allowing extended families to live together. In the post-disaster rebuild effort, the government stepped in and built single storey brick and concrete houses, arranged in a grid formation. The houses had large windows to let in a lot of light and were allocated by drawing lots. This allocation process meant extended families were no longer able to live together, the fact that the houses were single storey meant there was nowhere to house livestock, and the large windows failed to provide appropriate levels of privacy. Thus, the new housing did not allow the families of the region to practise their primary occupation or their religious customs. As a result, the housing was rejected by some, who chose instead to gather vernacular materials and rebuild their own house in the vernacular style. Others accepted the government-provided housing, and then began to build around to recreate the courtyard centred style they were accustomed to, eventually allowing extended family units to move in.

This highlights two points to consider in post-disaster rebuilding. Firstly, the importance of understanding the style of housing that existed pre-disaster. Secondly, the fact that many people rejected the government housing and built a vernacular structure on their own highlights their capability. By engaging with the local community and using the existing resources in terms of building labour the government could have saved a lot of money and provided a more appropriate solution.

Nepal

On 25 April 2015 a magnitude 7.8 earthquake shook the region of Gorkha, Nepal, followed by a 7.3 magnitude aftershock 17 days later (Reid, 2018a). More than 600,000 houses were destroyed and a further 228,000 were damaged (Rafferty, 2019). The earthquakes occurred just before the country's monsoon season, severely hindering rebuild efforts in the region. In the post-disaster rebuild effort, the traditional vernacular structures that once dominated the region were rejected in favour of modern brick and concrete houses. These modern house designs were considered superior to vernacular structures in terms of earthquake resilience, and this was naturally the focus of the rebuild effort. Unfortunately, the modern houses failed to cater for the extreme weather conditions experienced here such as the freezing winters, high winds and monsoons, which the vernacular houses had been designed to cope with.

The assumption that the modern house designs had superior earthquake resilience was also questioned. Forbes (2018) studied surviving houses after the disaster and noted that even though many vernacular structures had been destroyed, there were some still standing intact. There were also many modern style houses that had been destroyed in the earthquakes. Thus, the authors concluded that building earthquake resilience was not so clear cut. Deeper investigation revealed that some of the vernacular style houses were built after the previous severe earthquake experienced in the region in 1934. At that time, the wood traditionally used for the timber frames was in short supply. As a result, many people had harvested immature trees to build with, which did not have sufficient strength to cope with the shock of the earthquake. Replacing the timber frames with another material such as polypropylene geogrid bands or galvanised wire gabion bands could have also been a way to create a more earthquake resilient vernacular style building not considered during the rebuild effort.

Another issue that came up during the rebuild effort was that the government only pledged to build new houses to replace the ones destroyed by the earthquake (Forbes, 2018). No assistance was offered to repair houses that were damaged. Repairing a damaged house is usually much more economical than building from scratch, and ensuring houses are in good repair is important for their resilience to potential future shocks as well as ensuring they are safe and comfortable to live in. Thus, people in damaged houses were stuck with either continuing to reside in the damaged house or destroying their home so they could obtain funding to rebuild from scratch.

Mexico

In 2017 Mexico suffered two severe earthquakes in close succession, one of magnitude 8.1 on 7 September in Southern Mexico, followed by one of magnitude 7.1 on 19 September in central Mexico (Huber, Klinger, and O'Hara, 2017). The combined impacts saw an estimated 184,000 houses damaged by the earthquakes (Reliefweb, 2017). The earthquakes affected houses in both urban areas including Mexico City, and poor rural areas in the southern states of Chiapas, Guerrero and Oaxaca (The Conversation, 2017).

The rural communities were worse affected as they lost not only their homes but their agricultural assets, their main sources of income. Multiplying impacts saw these families, who were already living in poverty, struggle for survival. The loss of crops meant the families were affected immediately as this was their main source of sustenance. It also meant that the amount of food available in the region was in short supply, which led to an increase in food prices. Thus, these communities were left with no homes, no source of income and decreased food supplies. It was reported that government funded rebuild efforts focused on the more densely populated urban areas and that many rural communities did not receive any government support at all during the initial aftermath of the earthquakes (Neuhausen, 2018). Instead, self-organised groups of volunteers travelled to the devastated rural communities with donated supplies to aid the rebuild efforts. The groups were largely organised through social media, where they advertised for skilled labour and supplies. The groups were respectful of the many indigenous cultures in the target communities and ensured their needs were catered to. This highlights the important role non-governmental organisations can take in leading rebuild efforts, especially in countries with high levels of poverty.

Tsunamis

Tsunamis, also known as seismic sea waves, and formerly referred to as tidal waves, occur when a huge surge of ocean water, potentially reaching over 30 metres high, crashes onto land causing severe damage to coastal areas. Tsunamis are often associated with the aftermath of an earthquake, but they can also be caused by tectonic plate shifts, volcanic eruptions, underwater landslides and even meteorites.

The worst tsunami in recent history arose in the wake of a 9.1 magnitude earthquake that shook northern Sumatra on 26 December 2004. The resultant seismic wave saw huge surges of ocean water rapidly moving toward coastal areas of 14 countries throughout the Indian Ocean region. Impacts were reportedly experienced as far away as North America and Antarctica. It is estimated that the death toll from the tsunami exceeded 230,000 people. Those that survived had a massive clean-up and rebuild as thousands of buildings were also severely damaged. The number of buildings damaged in some of the worst affected countries includes 1300 in Myanmar, 1500 in Malaysia, 4800 in Thailand and 6000 in the Maldives. However, much more significant damage was inflicted on Sri Lanka where 119, 562 buildings were damaged, India where 157, 393 were damaged and in Indonesia where the figure was 179, 312 (ABC News, 2014).

The global response to the disaster was extremely generous and the resultant coordinated efforts of the United Nations in the aftermath of the tsunami were relatively successful. For instance, there were no serious outbreaks of infectious disease, as usually occur when thousands of displaced people are forced into temporary shelters. This is largely attributed to exceptional management of water supplies. Other successes were the emphasis placed on improving opportunities for the many children who had been orphaned by the tsunami and increasing opportunities for women. Thus, the "build back better" philosophy was applied to communities as a whole, rather than just construction (United Nations, 2009).

Despite the comparatively successful response in the immediate aftermath of the disaster, the rebuilding was slow with only 20 per cent of displaced people reported to have been returned to permanent housing one year after. Land tenure issues were one reason cited as causing significant delays. People's land ownership documents had been washed away with the buildings and so some people had difficulty proving their right to land ownership. The massive death toll also meant many landowners were gone, and the decision about who should have the right to rebuild on the vacated land was heavily contested in some cases (Hays, 2008). The emphasis on improving resilience to future disasters also caused some friction. The only real way to prevent tsunami damage in the future was to relocate houses in the most vulnerable parts of the coastal areas further inland; however, this was contested for various reasons. For example, some of the coastal communities rely on fishing for sustenance and income and thus living away from the coast would impact their livelihood.

Volunteer organisations that came in to aid rebuilding efforts also created some conflict. Firstly, the many areas affected by the disaster comprised many different religions and ethnicities. Some of the volunteers came as part of religious organisations that were seen as taking advantage of the situation to proselytise. They were noted to have handed out religious dogma along-side the food supplies, disrespecting the local culture and in some areas sparking conflicts (Hays, 2008). Local construction workers were also upset by some of the volunteer builders. They may have found it difficult to obtain paid work while volunteers were coming in to do their jobs for free. This highlights the importance of respecting the local community during rebuild efforts and maximising the utilisation of local resources, only bringing outsiders to work on rebuilding were local resources are unavailable.

Cyclones

Severe storms have various names; they are known as hurricanes in the north-east Pacific and north Atlantic, named after the Caribbean god of evil; in the northwest Pacific they are called typhoons, which is believed to derive from the Urdu word for violent storm. In the South Pacific and Indian Ocean regions they are referred to as cyclones, tropical cyclones or cyclonic storms and other variations. The word cyclone derives from moving in a circle (Wikipedia, 2019c). For the remainder of this book we refer to them as cyclones.

Cyclones are created by the evaporation of relatively warm ocean water recondensing as clouds. Thus, they form over the warmer tropical seas impacting any nearby land masses. The temperature of the oceans is rising, believed to be due to accumulation of greenhouse gases in the atmosphere, which is impacting the formation of cyclones. They are becoming less frequent but more intense (Lefale, Diamond, and Anderson, 2018). Countries located in the tropics are used to frequent cyclones, however, due to their increased severity in recent years, the impacts have been devastating. In particular, the small Pacific Island nations have been some of the worst hit. Here, we look at the impacts on housing of some of these.

Cyclone Pam

Cyclone Pam, rated category 5, tore through the Pacific in March 2015 affecting several of the small island nations but perhaps worst affected was Vanuatu. Extensive damage to buildings saw 75,000 people left homeless and the cyclone tore through agricultural lands devastating around 96 per cent of crops (Bolitho, 2015). Amazingly only 11 lives were lost during the storm, believed to be the result of local knowledge in how to survive cyclones due to their frequency of occurrence in the region (Connors, 2016).

The rebuilding process was made possible through the aid of international governments and charitable organisations. In efforts to build back better many new approaches were taken to improve resilience. For example, prior to the cyclone, most houses had a stove for cooking that used indoor biomass burning, which was linked to the widespread respiratory illnesses in the region. In the build back initiative, outdoor communal kitchens replaced the indoor ones, making healthier indoor environments (Reid, 2018b).

Habitat for Humanity also had strong involvement in the rebuild programme. Their approach was to operate the rebuilding as a training programme, teaching the local community construction skills that create more resilient buildings, not just for potential future cyclones, but other shocks that affect the region, which is in close proximity to an active volcano. Along with the training, the programme provided the local community with a bank of tools such as saws and hammers. Thus, rather than simply rebuilding the community, the programme supplied them with the ability to rebuild their own homes, reducing their dependence on aid when disaster strikes again (Habitat Australia, 2016). This is an important approach for small nations in the Pacific region, due to the frequency of cyclones in the area. If rebuild takes more than a year, the next cyclone season will be upon the community before they are back in permanent housing, increasing their vulnerability. Giving communities the ability to rebuild for themselves will make the process much more efficient.

Cyclone Winston

In 2016, a year after the cyclone Pam disaster, cyclone Winston tore through the Pacific Islands, and this time Fiji caught the brunt of the devastation with 40,000 homes destroyed (Wikipedia, 2019b) and 131,000 people left homeless (World Bank, 2017b). The rebuilding process in Fiji was slow, particularly in densely populated urban areas with many displaced people in temporary shelters and damaged houses still awaiting repairs six months after the disaster. The shortage of building materials was the main reason for delays in the rebuild effort (Fox, 2016). While this may seem like a short amount of time compared to other rebuild efforts, such as those described in the earthquake and tsunami sections, it meant that the new cyclone season was around the corner, and the partially damaged houses were likely to be impacted even if the cyclones were less intense next season.

In the rural villages, however, a different situation emerged. These small close-knit communities showed great initiative in coming together and developing their own rebuilding efforts. For example, in the province of Ra, one of the most ethnically and linguistically diverse regions in Fiji, communities worked together to rebuild each other's homes, as well as communal village buildings such as schools (World Bank, 2017b). The villages are described as the kind of place where if you see anyone while you are out you always say hello. People are still struggling to rebuild not just their houses, but their lives, as for many their source of income was lost in the cyclone as well. The local industry was largely based on coconuts, raising pigs and tending beehives, much of which were destroyed by the cyclone – trees and beehives were washed away and many pigs were killed. But the strong sense of caring the community has for each other is helping the people to thrive (World Bank, 2017a). This highlights how living in a close-knit rural community in itself is a form of resilience to disaster.

Hurricane Gita

The next major cyclone in the Pacific, Cyclone Gita, tore through the region in February 2018. This time Tonga was the worst affected nation with minor damage caused to 5499 properties, major damage to 1028 properties and complete destruction of 469 homes (Matangi Tonga, 2018). Initially, many of the displaced people were housed in canvas tents. Then structures consisting of a concrete floor, wooden frame and corrugated iron roof were erected. Efforts to add walls to these temporary structures are underway to finally see residents returned to a permanent structure.

CARE Australia has been involved in the rebuild efforts. Like other rebuilding programmes, they have tried to improve communities through their rebuilding efforts. Some of the initiatives include improving hygiene and sanitation facilities and being more inclusive of women and marginalised groups (Australian Humanitarian Partnership, 2018). They have also assisted in replanting efforts to recover the agricultural assets lost during the disaster.

Cyclone Fani

Cyclone Fani hit the Indian state of Odisha in May 2019. Fani was formed west of Sumatra in the Indian Ocean and rapidly intensified into an extremely severe cyclonic storm which hit the Indian East Coast and then the South of Bangladesh. Cyclone Fani resulted in relocation of over a million people across Odisha and part of Bangladesh. While the pro-active authorities have managed to contain the loss of lives to just seven in India, overall combined damage to properties in India and Bangladesh was in excess of US$1.81 billion. Many public buildings including schools, hospitals and office buildings were damaged due to sub-standard construction and poor maintenance (Gupta, Regan, and Berlinger, 2019).

Floods

The basic definition of flooding is water accumulating where it is not wanted (Australian Government, 2019). While flooding is most commonly associated with periods of heavy rainfall, floods can occur in dry areas if they are downstream of a water course and various other causes. While flooding tends to be the result of water courses such as lakes and rivers failing to contain rainfall, the impacts of flooding can be exacerbated by anthropogenic influences. For example, alterations to the landscape make it less absorbent, causing water to accumulate at ground level. These alterations include concreting over parkland; even altering the type of plant species in an area can cause the soil to become more compacted reducing its absorbency. While flood waters in themselves create widespread devastation, they can also generate side effects such as mudslides that can bury whole villages, destroying homes and agricultural plantations.

In much of Southeast Asia, particularly India the seasons are dictated by monsoons (National Geographic, 2019b). In the summer these bring torrential rainfalls. The region has developed around these seasonal periods and relies on the heavy rainfall in many ways. For example, many agricultural properties have no formal irrigation systems and rely on the rainfall for crop growth and animal management. A lot of people rely on wells for their water supply, and the heavy rainfall fills up the groundwater sustaining this supply. Much of the country's electricity supply relies in hydropower. When rainfall is lower than expected the economy can be negatively impacted as these essential resources may be in short supply, however, when rainfall is heavier than expected the situation can be far worse. Widespread flooding can destroy properties and damage possessions. Many people are left with no home and no way to generate an income as their agricultural resources can also be lost.

In 2018 the monsoon season brought particularly heavy rains throughout India leading to widespread devastation. The state of Mizoram was first affected as landslides brought on by the heavy rainfall resulted in several casualties (Reliefweb, 2018). By the end of June flooding had already affected at least one million people across the country in the states of Assam, Tripura, Manipur, West Bengal, Maharashtra and Kerala. The rains continued, and many states continued to be affected, but by far the worst affected was Kerala. According to reports, more than 700,000 people were displaced in the state (Dahshan, 2018). The state is home to significant mining and hydropower plants and these were believed to have exacerbated the impacts of the flooding. Reports claim that the hydropower dams were dangerously close to overflowing when managers finally decided to release the water, contributing significantly to the flood related damage (Reliefweb, 2018). If releasing the waters had begun earlier and been done more slowly it is argued that much of the damage may have been prevented.

In the immediate aftermath of the disaster, once the flood waters finally receded, the streets were filled with waste and clearing this up posed the first challenge, before even considering rebuilding (Chauhan and Raghuram, 2018). The content of the waste was likely to include many hazardous substances, including carcasses, and needed to be dealt with carefully to prevent the outbreak of disease. The rebuild effort is still ongoing. Dahshan (2018) highlights six important considerations that should form part of the reconstruction efforts to ensure longevity and resilience against future disasters. These considerations resonate with many of the themes presented throughout this book including: 1) the importance of approaching rebuilding holistically. If one department focuses on housing, another on transport, another on water, the result will be disjointed and cumbersome. Instead, a coordinated approach should be used; 2) using the rebuild as an opportunity to improve the quality of buildings in the region; 3) engaging the local communities in the rebuild effort and ensuring that all peoples, irrespective of ethnicity or religion are given equal support; 4) prioritising the resumption of income generating activities to ensure the recovery of the economy; 5) taking advantage of the donation window, when international interest in the disaster will create a peak in international donations; and 6) integrating environmental sustainability into the rebuild. While the level of disaster appears unusual, the impacts of climate change project that such disasters will occur more frequently, and so ensuring better resilience against future heavy monsoon seasons should be a priority. This must go beyond simply improving the quality of buildings to improving the overall landscape.

Drought

While the natural disasters considered so far have focused on sudden short-term events, slower on-set droughts can be equally devastating. The 2011–2012 droughts in East Africa for example resulted in the deaths of 260,000 people (Wikipedia, 2019a). The causes of death varied from malnutrition, which was brought about by the combined inability of the dry soils to support plant growth and the loss of livestock to dehydration and starvation. This loss of agricultural produce also resulted in the loss of livelihood for many of the residents in the region who depend on agricultural activities. Another significant cause of death was related to infectious diseases brought on by the lack of sanitation due to the lack of water resources, coupled with overcrowding conditions as families fled to refugee camps seeking aid.

The impact of droughts on housing is less direct than the other natural disaster described, as the houses are not destroyed. However, the affordability of housing is impacted, as families lose their source of income and can no longer afford to pay accommodation costs. Many families will be forced to move to try and find a property in a region with more fertile soils, also contributing to housing difficulties.

Drought impacts in rural communities can be exacerbated by preference for urban areas and industries, even where rainfall is not unusually low. For example, in India the drive to boost the economy has favoured the increase of output from industry at the expense of rural villages. Walker (2008) reports in Gujarat in 2006 the amount of water diverted to industry from rural areas increased five-fold, creating drought conditions for many rural villages. Groundwater resources, essential to the survival of many rural communities that rely on wells for their water supply, have reportedly been extracted by soft-drink companies diverting it away from local communities and leaving them with no drinking water, no water for irrigating their crops, no water for washing, leading to outbreaks of communicable diseases and other devastating health effects.

Wildfires

On top of the negative impacts droughts can have on rural communities discussed above, they also create optimal conditions for wildfires. Dry vegetation provides plenty of fuel for a fire to burn out of control, while nature provides the oxygen and heat sources needed to ignite the fires. The extremely dry conditions mean fires can also easily be ignited accidently through dropping a lit cigarette, or biomass burning for cooking (National Geographic, 2019a). Fires also commonly occur after an earthquake due to the combination of burst gas pipes releasing flammable gases into the atmosphere and electricity sparks igniting the gas.

In some developing economies, slash and burn agriculture is still widely practised. This usually occurs in forested areas, where the vegetation will be razed and set on fire. This creates a layer of fertile ash, which is planted with crops. Unfortunately, the method only creates a temporary fertility and after one or two seasons the soils are depleted of nutrients and the nomadic communities move on to another area. While this practice has been ongoing for many years in some cultures, newer conditions have made it more dangerous. For example, climate change is making forest areas drier than ever before, and the burning of the razed field can easily get out of control, sending flames into the extended forest area. Similarly, as the population continues to expand, rural communities are living in closer proximity to forested areas, and therefore the slash and burn fields are coming closer to permanent villages putting their housing at risk.

One of the most devastating wildfire outbreaks occurred in Indonesia in 1997–1998. The fires were reportedly due to slash and burn agricultural activities in an area that had suffered a season of drought. The fires burned an area of 19,768,430 acres and resulted in 100,000 casualties. However, other analysts blame the many industries present in the region who convert the forest area into palm oil plantations and other controversial pursuits (Gellert, 1998). While the fires mainly impacted Indonesia, neighbouring countries were severely affected by the smoke including Thailand, Malaysia, Brunei, Vietnam, Singapore and the Philippines.

Conflict

The disasters described so far in this chapter have partially natural causes, with some anthropogenic influences. A more directly anthropogenic form of disaster is conflict. Conflict can destroy whole villages, not just their physical attributes like buildings and infrastructure, but their community spirit. We live in a globalised world, which is making us more aware than ever of the social, political and cultural diversity that exists. For a close-knit rural community to thrive, these differences must be respected and celebrated, to avoid destructive outbreaks of conflict. Like natural disasters, post-conflict rebuilding must build back better, however, in this instance the philosophy needs to go beyond physical structures to create peaceful communities. As defined by the Carnegie Corporation (1997), the aim of preventive action is to prevent the emergence of violent conflict, prevent ongoing conflicts from spreading and prevent the re-emergence of violence.

Conflicts can carry on for many years creating some unique challenges during post-disaster rebuilding. One issue is when to start rebuilding. In many places conflict may cease in a particular area, even though it is still going on in other parts of the country (Barakat, 2003). People who are trying to get on with their lives in the peaceful region need to start rebuilding straight away. However, it may be difficult for aid workers to assist in rebuilding, as concerns over their safety may prevent them from travelling to conflict affected areas. Due to the longevity of conflicts many young people may have lived their whole lives in conflict affected regions. As a result, they may have had no opportunities for formal education. Thus, engaging them in the rebuild could be a way to train them with practical skills they can use to generate an income in the post-conflict society. Another side effect of conflicts is the loss of life, which is likely to disproportionately affect young males who are preferentially recruited to participate in the violence. Thus, there may be many female-led households trying to rebuild. Programmes focusing on assisting women to rebuild and learn income generating skills will also be essential to their future well-being. Overall it is important that people are not just working to rebuild their own homes, but that the community is working together to rebuild houses for all.

In Sri Lanka, where long term conflict afflicted the nation from 1986 to 2009, the loss of housing was severely exacerbated by the 2004 tsunami which destroyed more than 95,000 houses (Haigh et al., 2016). The combined impact of the tsunami and the conflict saw 280,000 people displaced when conflict finally ceased in 2009. Haigh et al. (2016) studied four different post-disaster housing projects in the region. In three of the regions studied, allocation of housing had been biased towards majority groups. The arrangement of housing allocation discouraged community integration, with people of different religious or ethnic identities being housed in segregated areas. These segregations were often accompanied by a bias, that is, the minority group locations were placed in a way that made it difficult for them to access public amenities and services. As a result, the cultural differences that fuelled the conflict are being kept alive and outbreaks of violence have been reported in these areas. In contrast, the fourth area had allocated housing preferentially for low-income households, female-led households and other minorities, while also ensuring the size of housing was appropriate for the number of household members. This region has seen no violence since the conflict ended, indicating that improving equality has a strong hand to play lessening and preventing conflict.

Conclusion

When a significant disaster occurs, worldwide news will report on the impacts for a short period of time until another more interesting story comes along. Despite the fact that post-disaster rebuilding efforts take years, it is during this short time, usually two to three months after the disaster is first reported, that the global community will empathise with the disaster affected people and wish to contribute to rebuilding efforts (Habitat for Humanity, 2019). Thus, it is important that relief aid agencies take advantage of this window and advertise for people to make donations. While it is great that there are many charitable organisations around, these often work independently of each other. Thus, people will be making donations to many different organisations that serve a different role in rebuilding efforts. Unfortunately, as a result, multiple organisations may overlap in the type of post-disaster assistance they offer or may offer inappropriate assistance. For example, a shipment of food donations sent to a Pacific Island nation had to be turned away, because the produce did not comply with airport customs (Habitat for Humanity, 2019). Instead, charitable organisations should consult with the government of the affected region to determine what types of donations are required and acceptable. Different charitable organisations should work together to ensure they are not doubling up on support but providing a coordinated and effective relief effort.

After many disasters NGOs organise groups of volunteers to build replacement housing in the affected area. Due to the limited building experience of these volunteers, there is very little flexibility in the designs of the replacement houses, and thus it is difficult to be accommodating to different families' lifestyles. Many regions may have capable builders and construction workers living in their midst, however, these skills are often overlooked in the influx of volunteering efforts. A better approach to volunteer building is working closely with the target community. Firstly, it is important to assess the level of skill available on the ground and engage local builders in the projects. Secondly, allowing input from the local community into the building design can ensure housing is appropriate. Thirdly, for unskilled locals, engaging them in the design and building process can help them develop new skills, allowing them to maintain and repair their own homes going forward and thereby increasing disaster resilience.

After a disaster, there are conflicting priorities to rebuilding. On the one hand, many people have lost their home and their source of income and thus, rebuilding quickly is a priority for survival. On the other hand, there is a desire to "build back better", ensuring the new buildings are more resilient. The build back better theory often goes hand in hand with modern building materials such as brick and concrete. Other researchers believe that the most logical approach would be to study the buildings that are still standing after the disaster and emulate their resilience features. Unfortunately, this research would be time consuming and time is not a luxury that can be afforded in the wake of a disaster. For many regions, disasters are not a freak occurrence but a fact of life: for example, the hurricanes in the Pacific Islands and flooding in India. Thus, instead of waiting for disaster to strike and applying the build back better approach in the aftermath, work should be going on now to ensure rural communities are prepared to cope when disaster strikes. This should entail building a database of disaster resilient design features and implementing this resilience building in regions that are most vulnerable. In the following chapter we will look at design features that can improve resilience for some of the disasters discussed.

References

ABC News. (2014). Boxing day tsunami: How the disaster unfolded 10 years ago. Retrieved from www.abc.net.au/news/2014-12-24/boxing-day-tsunami-how-the-disaster-unfolded/5977568

Ambraseys, N. N., and Tchalenko, J. S. (1970). The Gediz (Turkey) earthquake of March 28, 1970. *Nature*, 227, 592–593.

Australian Government. (2019). Flood. Retrieved from www.ga.gov.au/scientific-topics/hazards/flood

Australian Humanitarian Partnership. (2018). Tropical cyclone Gita – Tonga. Retrieved from www.australianhumanitarianpartnership.org/preparedness-1/tropical-cyclone-gita

Barakat, Sultan. (2003). *Housing reconstruction after conflict and disaster*. London: Overseas Development Institute.

Bolitho, S. (2015). Tropical cyclone Pam: Why was the Vanuatu death toll so low? Retrieved from www.abc.net.au/news/2015-04-01/explainer3a-why-was-the-vanuatu-death-toll-from-cyclone-pam-so/6363970

Carnegie Corporation. (1997). *Preventing deadly conflict: Final report*. New York: Carnegie Corporation.

Chauhan, S., and Raghuram, S. (2018). Rebuilding life in Kerala after the floods. Retrieved from www.forbesindia.com/blog/infrastructure/rebuilding-life-in-kerala-after-the-floods/

Connors, A. (2016). Cyclone Pam: Vanuatu one year on. Retrieved from www.abc.net.au/news/2016-03-13/cyclone-pam-vanuatu-one-year-on/7242620

The Conversation. (2017). Mexico's road to recovery after quakes is far longer than it looks. Retrieved from http://theconversation.com/mexicos-road-to-recovery-after-quakes-is-far-longer-than-it-looks-84479

Cromley, E. (2008). Cultural embeddedness in vernacular architecture. *Building Research and Information*, 36(3), 301–304.

Dahshan, M. E. (2018). Six lessons for the sustainable reconstruction of Kerala. Retrieved from www.orfonline.org/expert-speak/43798-six-lessons-for-the-sustaina ble-reconstruction-of-kerala/

Forbes, C. (2018). Rebuilding Nepal: Traditional and modern approaches, building or diminishing resilience? *International Journal of Disaster Resilience*, 9(3), 218–229.

Fox, L. (2016). Cyclone Winston: Fiji struggles to rebuild six months after storm devastated the country. Retrieved from www.abc.net.au/news/2016-08-26/fiji-s truggles-to-rebuild-after-cyclone-winston/7789720

Gellert, P. K. (1998). A brief history and analysis of Indonesia's forest fire crisis. *Indonesia*, 65, 63.

Gupta, S., Regan, H., and Berlinger, J. (2019). 7 killed as tropical cyclone Fani hits India. Retrieved from https://edition.cnn.com/2019/05/03/asia/india-landfall-cyclo ne-fani-wxc-intl/index.html

Habitat Australia. (2016). Reflections from Vanuatu – rebuilding lives after cyclone Pam. Retrieved from https://habitat.org.au/reflections-from-the-field-rebuilding-a fter-pam/

Habitat for Humanity. (2019) Pacific Peoples Housing Forum, 17 May, 2019, Auckland, New Zealand.

Haigh, R., Hettige, S., Sakalasuriya, M., Vickneswaran, G., and Weerasena, L. N. (2016). A study of housing reconstruction and social cohesion among conflict and tsunami affected communities in Sri Lanka. *Disaster Prevention and Management: An International Journal*, 25(5), 565–580.

Hays, J. (2008). Relief organizations, money and the December 2004 tsunami. Retrieved from http://factsanddetails.com/asian/cat63/sub411/item2548.html

Huber, C., Klinger, H., and Kristy J. O'Hara. (2017). 2017 Mexico earthquake: Facts, FAQs and how to help. Retrieved from www.worldvision.org/disaster-relief-news-stories/ 2017-mexico-earthquakes-facts

Lefale, P. F., Diamond, H. J., and Anderson, C. L. (2018). Effects of climate change on extreme events relevant to the Pacific Islands. *Science Review*, 50–73.

Matangi Tonga. (2018). Cyclone Gita cost Tonga $356 million. Retrieved from http s://matangitonga.to/2018/05/15/cyclone-gita-cost-tonga-356-million

National Geographic. (2019a). Climate 101: Wildfires. Retrieved from https://www.na tionalgeographic.com/environment/natural-disasters/wildfires/

National Geographic. (2019b). Monsoon. Retrieved from https://www.nationalgeo graphic.org/encyclopedia/monsoon/

Neuhausen, E. (2018). Rebuilding Mexico. Retrieved from https://nacla.org/news/ 2018/07/10/rebuilding-mexico

Rafferty, J. P. (2019). Nepal earthquake of 2015. Retrieved from www.britannica.com/ topic/Nepal-earthquake-of-2015

Reid, K. (2018a). Nepal earthquake: Facts, FAQs and how to help. Retrieved from www.worldvision.org/disaster-relief-news-stories/2015-nepal-earthquake-facts

Reid, K. (2018b). Cyclone Pam: Facts, FAQs and how to help. Retrieved from www. worldvision.org/disaster-relief-news-stories/cyclone-pam-facts

Reliefweb. (2017). Mexico: Earthquakes. Retrieved from https://reliefweb.int/disaster/ eq-2017-000138-mex

Reliefweb. (2018). India: Floods and landslides. Retrieved from https://reliefweb.int/ disaster/fl-2018-000134-ind

United Nations. (2009). Five years after Indian Ocean tsunami, affected nations rebuilding better - UN. *UN News*. Retrieved from https://news.un.org/en/story/2009/12/325422-five-years-after-indian-ocean-tsunami-affected-na tions-rebuilding-better-un

Wald, L. (2019). The science of earthquakes. Earthquake Hazards Program. Retrieved from https://earthquake.usgs.gov/learn/kids/eqscience.php

Walker, K. L. M. (2008). Neoliberalism on the ground in rural India: Predatory growth, agrarian crisis, internal colonization, and the intensifictaion of class struggle. *The Journal of Peasant Studies*, 35(4), 557–620.

Wikipedia. (2019a). East Africa drought. Retrieved from https://en.wikipedia.org/wiki/2011_East_Africa_drought

Wikipedia. (2019b). Cyclone Winston. Retrieved from https://en.wikipedia.org/wiki/Cyclone_Winston

Wikipedia. (2019c). Tropical cyclone. Retrieved from https://en.wikipedia.org/wiki/Tropical_cyclone

World Bank. (2017a). Leading a family, and a community, through and beyond Tropical Cyclone Winston: Rai's story. Retrieved from www.worldbank.org/en/news/feature/2017/11/06/leading-a-family-and-a-community-through-and-beyond-tropical-cyclone-winston

World Bank. (2017b). Resilience and love in action: Rebuilding after cyclone Winston. Retrieved from www.worldbank.org/en/news/feature/2017/11/07/resilience-lo ve-in-action-rebuilding-after-cyclone-winston

9 Resilience in rural communities

Climate change mitigation has been the focus of world organisations and governments for decades now. As wealthy industrialised nations argue over the best way to reduce emissions and to maintain global temperature rise within 2° Celsius of pre-industrial levels (United Nations Framework Convention on Climate Change, 2019), developing nations are bearing the brunt of their high polluting lifestyles. As we saw in Chapter 8, while these regions have suffered natural disasters for centuries, the severity has increased rapidly in recent years causing widespread suffering (Lefale, Diamond, and Anderson, 2018). These communities need a shift in focus from mitigation to adaptation through building increased resilience.

Looking at vernacular architecture, it can be seen that many rural villages have already incorporated resilience features into their houses. In many villages along the banks of the Brahmaputra River in India vernacular houses are built on stilts to cope with the seasonal flooding. Villages in Pacific Island communities such as Fiji and Vanuatu often have a central community building with a roof that slopes all the way to the ground, so that it cannot be blown away by the strong winds experienced during cyclones. Residents of these villages know to gather here during storms and warning systems are in place. However, in recent years the increased intensity of these events means some of the coping strategies in place are no longer adequate for communities to survive these events unscathed.

As we described in Chapter 8, post-disaster resilience building tends to adopt the motto "build back better" and indeed it makes sense that simply replacing what existed before will no longer be enough to cope with future disasters. The interpretation of better building tends to involve a move away from the vernacular architecture that had existed to modern brick and concrete structures. However, the appropriateness of this approach has been questioned. Firstly, these types of buildings often fail to provide adequate functionality for the lifestyles of their end users. Secondly, their perceived resilience may not be realised, depending on the types of disaster being faced in the region. Forbes (2018) and Solomons (2016) both note that while it is true that many houses have been destroyed or damaged during natural disasters, the style of houses that suffer damage is not uniform. Both modern and vernacular houses suffer damage, and both modern and vernacular houses remain intact. Forbes' (2018) investigation into the houses damaged during the 2015 Nepalese earthquake found that the houses most severely damaged were either poorly maintained and badly in need of repair; had had illegal additions made to them; or were inappropriately located. Thus, the extent of building resilience may be related to other factors than the building materials. Instead, studying the houses that remain intact post-disaster and determining the features that made them resilient is the only way to truly develop a resilient house.

Providing a community with inappropriate housing designs can pose a risk to the resilience of these buildings in various ways:

- If the layout of the housing does not suit the users, they are likely to attempt to make alterations which could make the buildings more dangerous, as was seen in Turkey (Cromley, 2008).
- If the local community are unfamiliar with brick/concrete building techniques they will not be able to perform essential repairs and maintenance to ensure the houses are in optimal condition, which may make them less resilient to disaster.
- Brick and concrete are generally not readily available in these regions, and must be imported from far away at great expense. This could prevent people from being able to perform repairs and maintenance on buildings in the future, even when they have the knowledge and skills to do so.

Figure 9.1 shows the interconnectedness of the issues associated with the development of housing and resilient communities. As the figure depicts, resilience of a community can be ensured by appropriate integration of the community-specific issues within the housing development programme. As the community grows, some of these issues may also change or new issues may emerge. A resilient community would require an adaptive approach to incorporate these issues in a time-phase manner.

The idea of spending time examining the surviving structures after a disaster to determine the ultimate in resilient building design is certainly logical. Unfortunately, the time it would take to complete this analysis is an unaffordable luxury in the wake of a disaster, when families are living in tents or other temporary structures, desperate to return to some semblance of normality. Communities need to be working now to increase resilience in their existing structures, before disaster strikes. As isolated rural villages increase their communication with the global community through increased access to media, awareness of their vulnerability is growing. Information about climate change impacts and how they are projected to impact specific regions is reaching these remote villages and motivating communities to develop adaptation schemes. This community level action has been praised for drawing on local knowledge and

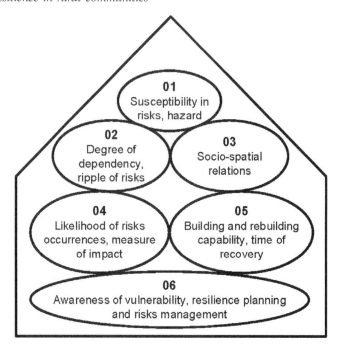

Figure 9.1 Development of housing and resilient community

resources, and maintaining cultural appropriateness. Many of these communities already have coping strategies in place which simply need to be scaled up to cope with the increasing severity of the natural disasters. As a result, these groups are attracting funding where they can show a clear and achievable resilience plan.

Critics of these community scale projects have expressed concern over their localised nature. These projects are often applicable at the village level and cannot easily be scaled up for implementation into the wider region (Ayers and Forsyth, 2009). All villages in these regions need improved resilience and allocating funds to projects that only help a few hundred people is seen as a misuse of resources when thousands need improvements. Thus, it is argued that resources would be more appropriately used to develop projects that can be implemented on a much broader scale.

In this chapter we look at design features developed to improve the resilience of buildings against specific disasters including earthquakes, tsunamis, cyclones, flooding and wildfires. We will also discuss some examples of implementing these, comparing preparedness approaches for different vulnerabilities.

Earthquakes

Earthquakes cause the ground to shake which can cause buildings to collapse. For residents, one of the biggest dangers aside from falling during the shaking is having parts of the building fall on them. Concrete and brick may be perceived as more resilient to collapse during an earthquake, however, if it does crack and collapse, its heavy weight means it poses a significantly greater risk to residents. On the other hand, lighter weight vernacular building materials such as bamboo and thatch may be perceived as more vulnerable to collapse in an earthquake, however, if they fall and land on residents the damage they inflict is far less severe, making them less of a hazard overall. There are many components of a building that can impact its earthquake resilience including the shape of the building, size of openings such as doors and windows, rigidity distribution, ductility, construction quality, roofs. In this section, however, we focus on the fundamental attributes including foundations, building materials and building location.

Whether or not a building collapses during an earthquake is strongly dependent on the building foundations. During the event the foundations of the house will experience extreme shear forces, which can lead to cracks, and other damage, or in extreme cases building collapse. Bamboo has been found to have high tolerance to shear forces compared to many other building materials (Roach, 1996). Wood has weak points such as knots and grain lines which can give way under stress. Concrete foundations can be inflexible and may crack, whereas bamboo foundations are flexible and will move with the earth. Professor Jules Janssen oversaw an affordable housing development in Costa Rica that employed bamboo as the main building material in the late 1980s. In 1991, the region was struck by a 7.5 magnitude earthquake and amazingly, while many concrete and brick houses lay in rubble, not one of the bamboo houses was damaged.

This example describes the high earthquake resilience of bamboo structures while they are still relatively new and in good condition. However, with time bamboo foundations may lose their strength either through exposure to moisture causing them to rot, or deterioration from pest infestations. The light gauge steel foundations described in Chapter 6 may provide a structure with the flexibility of bamboo that can withstand moisture exposure and will not attract pests, increasing the longevity of the building's resilience.

As the foundations of the building are impacted, the walls will begin to be affected and may crack and start to crumble. The building material used in the walls is therefore also an important consideration in ensuring resilience. Again, while brick and mortar walls are the go to for resilience design, other building materials have shown good earthquake resilience. The Assam type house described in Chapter 7 is known for its excellent seismic performance. Assam is an earthquake prone region, thus the evolution of this house design incorporated the need to withstand earthquakes. Kaushik and Babu (2009) summarise various earthquake events and how different Assam type houses were affected. They found that the houses with ekra walls, a type of grass commonly found in the region, were far more resilient than the concrete buildings and that risk of injury from the lightweight ekra buildings collapsing was significantly reduced compared to the collapse of concrete. Another innovative solution came from Kasapoglu (1989) who developed bricks that interlock together like LEGO pieces, removing the need for mortar bonding. These bricks have been shown to be able to withstand the shaking experienced during earthquakes better than traditional brick and mortar.

A secondary consideration in building earthquake resilience involves looking beyond the building materials to the location of the building itself. Earthquakes can cause landslides, settlement and soil liquefaction, all of which can impact houses on the affected land. The extent of the risk depends on the weight of the house and the type of soil it is built on. The risk increases the heavier the building and the sandier the soil. It is also important to work within the landscape of the region. In the hilly regions of Nepal, traditional vernacular structures were built into the sides of the mountains. In more recent times, this tradition had been lost and buildings were located close to the edges of hillsides. After the 2015 earthquake it was noted that the houses built close to the edge of the hills experienced greater damage than those built back into the hill in the vernacular style (Forbes, 2018). Thus, houses on sandier soils need stronger foundations, and in mountainous regions, building the house into the hillside, rather than on the edge will help create better resilience.

Another significant risk created by earthquakes is fires. These are usually caused by cracked gas pipelines leaking gas into the atmosphere, coupled with a spark from the electrical system igniting the gas. A tiny spark can cause a big fire; in fact, in the 1906 San Francisco earthquake 90 per cent of the damage was reportedly due to fires, rather than direct impacts of the shaking (Frantz, 2019). Many developing rural communities were not electrified in the past, and therefore this risk did not need to be a consideration in vernacular building design. However, there are massive benefits associated with having access to electricity (United Nations, 2019) and thus there is a huge push to bring electricity to the remotest of rural areas. This adds another layer of consideration for building earthquake resilience, as many of the vernacular building materials discussed, such as bamboo and ekra, are highly flammable.

Multistorey buildings are at increased risk of earthquake damage, compared to those lower to the ground. The lower floors need to provide a strong support for the upper levels to ensure resilience. Villages along the Brahmaputra River in Assam, India provide an interesting case study in resilience building as the region is prone to both earthquakes and flooding. The vernacular building style includes a bamboo foundation built on stilts to cope with the flooding. However, the frequent exposure to moisture will deteriorate the quality of the bamboo, and the raised floor level will increase vulnerability to earthquakes. Similarly, the frequent flooding has caused significant erosion, increasing the sandiness of the soils. Thus, it is likely that the most appropriate solution for this region will involve a combination of modern and vernacular structure. One solution, as described in Chapter 6, involves a steel frame, that is better equipped to withstand moisture, yet still able to withstand shear forces like bamboo. The walls can maintain the use of lightweight ekra to reduce the danger associated with collapse, as well as keeping costs low and improving sustainability.

A final consideration for building earthquake resilience is ensuring buildings are maintained in good condition. No matter how well a building is made initially, various effects over time can reduce its ability to withstand shocks. Especially for housing in earthquake prone regions, where a small shock may cause no visible damage to the building, there may be minor damage that will decrease the building's resilience when the next shock occurs. Thus, even after a minor shock buildings should be inspected and minor damage should be repaired. This was also noted in Nepal after the 2015 earthquake when many vernacular houses were severely damaged (Forbes, 2018). This reinforced the perception that modern brick and concrete houses are more resilient. However, the vernacular houses that survived the quake caused architects to reconsider and closer inspection revealed the damaged houses were in poor repair, for example had rotten wood, or unregulated building extensions. For an example of the success of routine maintenance, see the San Francisco church in Santiago, Chile (Jorquera et al., 2017). This church was built in the 1600s, and despite the many earthquakes experienced in the region over the years still stands today. Close inspection of the building shows repairs performed on the church periodically helped to maintain its earthquake resilience.

Tsunamis

Building resilience against tsunamis involves creating buildings capable of withstanding a huge influx of water. Thus, the design features largely overlap with those to be discussed in the section on flooding below. The main difference is that while there are many ways to divert water away from buildings during a flood, no such luxury is afforded during a tsunami. The only way to build true resilience is to build houses further inland, away from the vulnerable coastal areas. This situation is difficult because most people have an innate desire to rebuild their home in the same place. It brings up issues of land ownership, of maintaining the community spirit, and having access to the resources used for income generation. For example, in coastal areas fishing often forms a primary source of sustenance and income and having convenient access to the ocean will be a priority for locals. In Indonesia, after the 2004 disaster new building regulations were established, the core of which was placing restrictions on building locations (Stanley and

Seng, 2013). Monetary compensation was to be offered to locals allowing them to build elsewhere. Unfortunately, the lack of desire from locals to build in a new location, coupled with corruption in the Indonesian Government, meant the regulations were poorly enforced. More recently, the Indonesian Government has been cracking down on corruption and hopefully this will lead to improving resilience for coastal communities before the next tsunami inevitably occurs.

Buildings cannot offer shelter during a tsunami, the only way to survive is to run for higher ground. Fair warning of the approaching water is the most important factor for minimising the death toll from this form of disaster. The tsunami itself can provide a brief warning as the trough of a tsunami wave usually reaches the shore first, creating a vacuum effect sucking coastal water seaward (National Geographic, 2019). This sucking makes a loud sound and usually occurs approximately five minutes before the wall of water hits the shore. Teaching locals to recognise this sound and having an emergency action plan in place could save lives. This would entail mass evacuation to a pre-identified safe space at higher ground. The space should be equipped with emergency supplies allowing locals to survive for a few days until help arrives. A more formal earlier warning system, however, could greatly improve survival rates.

The Pacific Ocean region has had tsunami early warning systems in place since the late 1940s (Stanley and Seng, 2013). These were initially developed in 1949 in response to the devastating tsunami that hit the Aleutian Islands in 1946. Another severe tsunami that impacted Alaska in the 1960s saw a substantial upgrade to this system. Despite tsunamis occurring more frequently in the Indian Ocean, no such early warning system had been developed by 2004, contributing significantly to the astonishing death toll. The lack of early warning system appears to have mainly political roots. The Aleutian Islands are territory of the U.S., thus are comparatively well developed. The nations around the Indian Ocean region have lower development levels, and many villages lack access to basic infrastructure such as electricity and telecommunications, making an early warning system impossible. The nations all have different languages further confounding the development of a central early warning system. The political situation of the region has also had an impact. Many of the nations have suffered conflict both internally and with each other. However, since the shocking 2004 disaster, the situation has improved and it appears the development of an early warning system is underway.

Cyclones

Cyclones are severe storms whose main features are heavy rainfall and extremely strong winds. Building features that create resilience against the heavy rainfall associated with cyclones will overlap with those described in the flooding section below. Here, we focus on the impact of the strong winds. The strong winds can break the connections between the different components of a house (Vrolijks, 1998). Therefore, the simplest way to improve resilience is to improve the strength of these connections. One of the sights most commonly associated with cyclone incidents is whole roofs being lifted off a building and blown away. This creates a hazard to both people inside the building who are now exposed to the storm, and people outside who could be severely injured by the roof when it lands. In an even more extreme case, if the connection between the walls and the foundations are weak, the whole house can be lifted away by the winds. Thus, resilience to cyclones is not so much dependent on the type of building materials used in each house component, but rather the way these materials are connected. The use of modern building materials such as brick or concrete does not therefore automatically spell increased resilience. In fact, the use of these materials can increase the risk of injury, as if they are moved during the storm they could cause severe injury to residents, compared to lightweight vernacular building materials.

In post-disaster rebuilding, when components of a vernacular house are often replaced with modern building materials, it is important that the connecting components are reassessed. For example, thatched roofs are often replaced with corrugated iron sheets. These are heavier than thatch, and this increased weight needs to be factored in when determining the best way to secure the roof to the walls. If it is not securely attached, the corrugated roof will create a hazard if it comes loose during a subsequent cyclone.

Another benefit to the use of lightweight, locally sourced building materials is the simplicity of replacement. A thatched roof can be recreated from locally grown grasses, and does not require a high level of expertise. A tile or corrugated iron roof on the other hand will need to be imported from further away and assembled by skilled labour. In the wake of a disaster, when many houses need significant repairs and rebuilding simultaneously, readily available building materials are likely to be used preferentially to repair buildings in urban areas. Rural village communities will wait for help. Similarly, where the local community is already in the habit of constructing their own homes, the need to wait for skilled labour to become available to assist with building in their region will also be avoided.

It may be difficult to build entire villages of cyclone resilient housing, particularly in areas such as the Pacific Islands where cyclones occur every year. For low resilience houses, staying inside could be equally as dangerous as being outside, therefore, in the interim, villages should prioritise building a resilient communal structure where citizens can gather for safety during a storm (The Conversation, 2015). Such buildings are a common feature of many Pacific Island villages and often involve a roof that slopes all the way to the ground, with strong connections. This prevents wind from getting under building components and splitting them apart as described above.

Floods

Flooding causes a range of impacts to housing. The damage can be super-ficial, as the house filling up with water ruins carpets and furniture. Over time, more serious structural damage can occur as the dampness causes rot to set in to building components and leads to the formation of mould. In cases of severe flooding whole buildings can be washed away instantly. Many regions such as the towns and villages along the Brahmaputra River experi-ence flooding incidents almost annually. Therefore, vernacular house designs in these regions have evolved in response to the necessity for flood resi-lience. Flood resilience measures broadly fall into two categories – resistant approaches, which incorporate measures that aim to prevent flood water from reaching buildings, and accommodating approaches which incorporate building designs that can cope with the stress of flood waters.

One common accommodating feature of vernacular flood resilience is raising the floor level to reduce the risk of water entering the living space of the house. Figure 9.2 shows a house on stilts and Figure 9.3 shows the Assam type house; while built on a concrete base note that the floor level is still elevated. Tradi-tionally, the stilt houses were built from bamboo, however, this limits the longevity of the buildings as over time the frequent exposure to moisture will result in rotting and these foundations will need to be replaced.

Figure 9.2 Vernacular stilt house

Figure 9.3 Assam type house

Another novel accommodating approach to flooding adaptation is building "floating houses". A vernacular type of floating house can be found around the Musi River in Palembang, Indonesia (Puspitasari et al., 2018). These houses are built on a flexible bamboo foundation that moves up and down with the tides. The frame of the house is usually made of wood, with doors and windows arranged to provide cross-ventilation for thermal comfort. The flexibility of this foundation also makes it resilient to strong winds. Another floating house design has been developed by architects from the BACA Bureau in the U.K. (Aleksić et al., 2016). This house has a reinforced concrete foundation with air pockets. As these air pockets fill with water during periods of heavy rain the floor level of the house will rise, and can move up to three metres from its resting surface. This could provide a superior alternative to building a house on stilts for regions such as Assam that suffer from both flooding and earthquakes. Being built on stilts reduces earthquake resilience.

Resistant approaches to flood resilience often look beyond the house structure to the surrounding landscape which can present important opportunities for rural communities. The soils surrounding the properties have the potential to absorb a lot of the water, and indeed are an important component of the water eventually subsiding. Therefore, maximising the absorbency of soils should be prioritised where possible. This has been noted in the U.K., where the conversion of riparian landscapes has led to a significant decrease in the absorbency and severely increased flood risk to properties in the surrounding area. Efforts are now underway to recreate the original landscape (Environment Agency, 2017). Similarly, at the Selah ranch in Texas, U.S. a massive regeneration project saw the dried-up aquifers refilled with water, as the replanting of native grasses improved the soil absorbency (Masters, 2017).

We have already described the heavy reliance developing rural communities have on agriculture. In flood prone regions, farmers have adapted to the conditions, and indeed rely on the periods of heavy rainfall for their crop management as no formal irrigation systems are in place. However, the combination of reduced soil quality due to monoculture agriculture, and increased severity of flooding is seeing these crops washed away in the heavy rains, leaving families with losses of food and income. Various approaches to improving agricultural resilience to flooding are also being developed. Firstly, better soil management can help sustain the integrity of the soils, coupled with growing plants that maximise soil absorbency around the crops to divert water away. Another innovative solution involves incorporating agriculture into the building design. Roof top gardens for example, can provide an alternative space for growing crops for essential sustenance in a more resilient location.

Bangladesh provides an important place for showcasing flood resilience. It is a flat, low-lying country mostly lying within 12 metres of sea level. It borders the Bay of Bengal, an arm of the Indian Ocean. Thus, its coastal communities are vulnerable to tsunamis as well as seasonal flooding from monsoons. Due to its low level it is also considered vulnerable to climate change induced sea level rises. Thus, many important flood resilience projects are being developed in the region. Ayers and Forsyth (2009) describe the development of floating gardens, which essentially comprise a soil-covered raft. Another innovative solution could be incorporating a roof top garden onto a floating house to provide both housing and agricultural resilience.

Another interesting case study area for the application of flood resilience strategies is East Boston in Massachusetts, U.S. Although this is an urban area, the resilience strategies are adaptable to rural areas. East Boston was built up around the coast as the docks popped up creating an area of employment for socio-economically disadvantaged members of the community. Subsequently, blocks of affordable apartments also appeared in the area. The industrial nature of the area and low socio-economic status of its residents have resulted in the area becoming derelict. In recent years flooding events have been occurring in the region and are projected to worsen. Zandvoort et al. (2019) explored ways to build resilience to the projected flooding in coming years. A series of "hard" and "soft" measures were proposed. Hard measures include building concrete walls to block the water from reaching the populated areas. Soft measures include re-greening pockets of land at the lowest points of the area to increase ground absorbency, thereby drawing water away from buildings. Re-establishing the wetland ecosystems close to the shore could also absorb a lot of water, preventing it from flowing further inland.

The benefits of hard measures are that they take up less space and allow more development in the area, for example increasing housing density to cater for the growing population and more industrial sites to increase employment opportunities. The benefits of soft measures are that they secure green space, which during the non-flood season can provide an amenity space for local residents, making the region more aesthetically pleasing and providing much needed green areas. While both measures have their pros and cons, the authors note that the most important aspect of the measures taken is to incorporate future projections related to climate change.

All the flood resilience measures described so far rely on the assumption that flood waters will eventually recede and things can go back to normal until the next event. However, the projected impacts of climate change include sea level rises (Church et al., 2013), which could see significant areas of land end up underwater permanently, particularly in low-lying coastal areas. Many of the adaptation projects conducted so far have been criticised for their failure to incorporate some of these future climate change projections. Instead, they divert valuable resources into a temporary stop-gap solution. The low education levels and lack of telecommunications equipment in developing rural regions means locals will not have access to the latest information relating to climate change projections. Therefore, community level projects will work under the assumption that events will continue to be manageable. It has been suggested that offering funding for community level adaptation strategies in vulnerable areas is a misuse of funds that would have been better spent relocating these communities further inland.

Another assumption implicit in the adaptation strategies considered here is that the building materials will be able to withstand the water quality during flooding events. While that may be true today, climate change is increasing the acidity of sea water which may inhibit the ability of common building materials to withstand exposure (Dobraszczyk, 2017). Concrete for example is corrodible by exposure to acid, although the acid resistance of concrete can be improved by incorporating additives into the concrete mix (Rao, Keerthi, and Vasam, 2018). However, trying to improve the properties of concrete may be putting a band aid on a broken leg. Concrete is not a sustainable building material, due to the high level of energy required in its manufacture and its limited supply. It is also expensive, and therefore not a practical solution for affordable housing. Investing in research into novel materials with superior qualities could be a better investment.

Biorock is a substance that was developed to aid restoration of coral reefs. It is formed by the electro-accumulation of minerals in sea water (Wikipedia, 2019). The basic principle behind biorock is that by running a mild electrical current through an underwater structure, the dissolved minerals calcium, magnesium and bicarbonate will adhere to that structure, acting like concrete. So far biorock has mainly been used as a protection or restoration process for coral reefs, but its high compressive strength and ability to renew itself with the addition of a mild electrical current, even in harsh water conditions, show its potential as a flood resilient building material particularly for coastal areas.

Another potential future building material is protocells. Protocells are synthetically manufactured single-cell organisms. Although artificially created they can reproduce and evolve much like bacteria and other single-celled organisms, meaning they are essentially alive (Bedau et al., 2009). Harnessing this ability in a building material means it could respond to stresses in its environment in a way that will keep it intact and over time may naturally adapt to its surrounding conditions. Although these novel building materials are still in the early stages of development, particularly protocells whose use as a building material is largely theoretical at this stage, they could form an important part of the future of resilient buildings. Concerns have been raised over the longer term impacts of these substances, especially protocells, whose convenient evolutionary properties may be unpredictable. However, it is likely that novel building materials are going to be an important part of the future.

Wildfires

Wildfires have multiple causes and are often the result of many things occurring simultaneously. Wildfires are an essential component of the lifecycle of forest areas, and attempting to prevent them from occurring altogether will actually lead to worse impacts when they do occur, as well as damaging the natural ecosystem of forests and the associated services they provide. The key to wildfire damage prevention is education as many developing rural communities use fire as part of their daily lives. Slash and burn agriculture involves the clearing of an area of forest of trees, then burning the remaining vegetation. Through this process a layer of nutrient rich ash remains on the soil creating a fertile soil for crop production. However, this is a temporary enrichment and after a year or two the soil nutrition will be depleted, and the farmers will move on to a new clearing. Although this method of agriculture has been practised for thousands of years, the expanding population has seen it lead to mass forest destruction, threatening biodiversity and contributing to greenhouse gas emissions. The increased drought conditions of many regions where slash and burn agriculture is practised is leading to these fires accidently burning out of control leading to loss of more forest area, and loss of buildings and infrastructure in nearby populated areas.

One of the key methods for reducing forest fires is therefore education – raising awareness among agrarian communities of the impact of the slash and burn activities, and how the impacts of climate change are leading to the threat of fires burning out of control. Attempting to increase understanding of ecological processes and support the farming of crops along with trees and promoting agroforestry will be essential to preventing wildfire devastation.

There are other options such as building houses out of fire resilient building materials, ensuring vegetation close to houses is fire-resistant. However, these options may be beyond the reach of many rural villages where these types of building materials are unavailable and unaffordable (EcoLogic Development Fund, 2019).

Defining resilience

In this chapter we have seen many methods proposed for improving the resilience of buildings. However, there is currently no standard against which to assess buildings to determine their resilience. Thus, while these design features will reportedly hold up against the next inevitable disaster, we won't really know until disaster strikes and we can see if the building is still intact. Several authors consider the need to establish a set of criteria that, if met will qualify a building as resilient. For example, Adamy and Bakar (2019) note the importance of re-establishing healthcare centres, especially hospitals. They develop a list of 15 criteria that must be incorporated into the design of a hospital post-disaster to ensure the building can withstand any subsequent disasters. To prove that these criteria will hold up in the face of a disaster, scale models should be created which can be tested in a laboratory setting.

Looking beyond housing all communities must have a safe space to gather in the immediate aftermath of a disaster. Many disasters come in multiple stages including earthquakes and tsunamis so once an initial incident has occurred, it is important to wait as further shocks may still be coming. We have discussed in previous chapters the importance of residents having access to community spaces for rural lifestyles. These communal spaces could double up as safe gathering places. Herman, Sbarcea, and Panagopoulos (2018) describe ways to adapt parks as temporary shelters during some types of disaster. For areas with flooding and tsunamis the safe gathering place must be located at higher ground and away from coastal areas. Care must be taken as sometimes trees can provide a natural screen for the debris that can come loose during disasters, however, in strong winds trees can be broken and add to the debris posing a serious hazard in themselves.

To make these areas truly safe spaces for gathering during a disaster, they should be equipped with essentials. Many people could be stuck there for days therefore it is important that there are cooking facilities. Many parks in developed countries will have BBQ facilities which could become life-saving in a disaster situation. Access to electricity would also be extremely helpful and ensure people could access communications equipment to both alert others to their need for help, as well as providing people with access to news of whether the disaster has passed, when it is safe to return to their homes.

Clean water is also an essential for human survival, thus fitting communal gathering areas with rain water storage tanks can provide both a source of potable water, and water for putting out fires depending on the type of disaster. A community garden as we have described previously has many societal benefits; in a disaster situation it could be the sole source of food for stranded residents for a few days and could thus become a life saving essential.

One of the most important factors in ensuring the success of a disaster relief park is ensuring that nearby residents are aware that they should gather here during the disaster and of all the features that have been incorporated into the park to aid their survival.

References

Adamy, A., and Bakar, A. H. A. (2019). Key criteria for post-reconstruction hospital building performance. *IOP Conference Series: Materials Science and Engineering*, 469. doi:10.1088/1757-899X/469/1/012072.

Aleksić, J., Kosanović, S., Tomanović, D., Grbić, M., and Murgul, V. (2016). Housing and climate change-related disasters: A study on architectural typology and practice. *Procedia Engineering*, 165, 869–875.

Ayers, J., and Forsyth, T. (2009). Community-based adaptation to climate change. *Environment: Science and Policy for Sustainable Development*, 51(4), 22–31.

Bedau, M. A., Durodie, B., Parke, E. C., Bennett, G., Caplan, A. L., Cranor, C. F., … Zoloth, L. (2009). *The ethics of protocells: Moral and social implications of creating life in the laboratory*. Boston, MA: MIT Press.

Church, J. A., Clark, P. U., Cazenave, A., Gregory, J. M., Jevrejeva, S., Levermann, A., … Unnikrishnan, A. S. (2013). Sea level change. In T. F. Stocker, D. Qin, G.-K. Plattner, M. Tignor, S. K. Allen, J. Boschung, A. Nauels, Y. Xia, V. Bex, and P. M. Midgley (eds), *Climate Change 2013: The physical science basis. Contribution of Working Group I to the Fifth Assessment Report of the Intergovernmental Panel on Climate Change*. Cambridge: Cambridge University Press.

The Conversation. (2015). Rebuilding a safer and stronger Vanuatu after cyclone Pam. Retrieved from https://theconversation.com/rebuilding-a-safer-and-stronger-vanuatu-after-cyclone-pam-42181

Cromley, E. (2008). Cultural embeddedness in vernacular architecture. *Building Research and Information*, 36(3), 301–304.

Dobraszczyk, P. (2017). Sunken cities: Climate change, urban futures and the imagination of submergence. *International Journal of Urban and Regional Research*, 868–887. doi:10.1111/1468-2427.12510

EcoLogic Development Fund. (2019). Slash and burn agriculture. Retrieved from www.ecologic.org/actions-issues/challenges/slash-burn-agriculture/

Forbes, C. (2018). Rebuilding Nepal: Traditional and modern approaches, building or diminishing resilience? *International Journal of Disaster Resilience*, 9(3), 218–229.

Frantz, C. (2019). The great 1906 San Francisco earthquake. Retrieved from www.infoplease.com/world/disasters/earthquakes/great-1906-san-francisco-earthquake

Herman, K., Sbarcea, M., and Panagopoulos, T. (2018). Creating green space sustainability through low-budget and upcycling strategies. *Sustainability*, 10(1857). doi:10.3390/su10061857.

Jorquera, N., Misseri, G., Palazzi, N., Rovero, L., and Tonietti, U. (2017). Structural characterization and seismic performance of San Francisco Church, the most ancient monument in Santiago, Chile. *International Journal of Architectural Heritage*, 11(8), 1061–1085.

Kasapoglu, K. E. (1989). Earthquake resistant brick design. In O. Ural and L. D. Shen (eds), *Affordable housing: A challenge for civil engineers*. New York: American Society of Civil Engineers.

Kaushik, H., and Babu, K. S. R. (2009). Housing report: Assam-type house. In *World Housing Encyclopedia*. Oakland, CA: Earthquake Engineering Research Institute.

Lefale, P. F., Diamond, H. J., and Anderson, C. L. (2018). Effects of climate change on extreme events relevant to the Pacific Islands. *Science Review*, 50–73.

Master, B. (Director) (2017). 50 years ago, this was a wasteland. He changed everything. YouTube: NationalGeographic. Retrieved from www.youtube.com/watch?v=ZSPkcpGmflE

National Geographic. (2019). Tsunamis 101. Retrieved from www.nationalgeographic.com/environment/natural-disasters/tsunamis/

Puspitasari, P., Kadri, T., Indartoyo, I., and Kusumawati, L. (2018). Microclimate and architectural tectonic: Vernacular floating house resilience in Seberang Ulu 1, Palembang. Earth and Environmental Science, 106 (The 4th International Seminar on Sustainable Urban Development).

Rao, K. J., Keerthi, K., and Vasam, S. (2018). Acid resistance of quaternary blended recycled aggregate concrete. *Case Studies in Construction Materials*, 8, 423–433.

Roach, M. (1996). The bamboo solution. *Discover*. Retrieved from http://discovermagazine.com/1996/jun/thebamboosolutio784

Solomons, M. (2016). Nev houses: Designer Nev Hyman creating flat-pack, cyclone-proof housing for vulnerable Pacific nations. Retrieved from www.abc.net.au/news/2016-04-19/nev-houses-surfboard-designer-flat-pack-cyclone-proof-housing/7335324

Stanley, D., and Seng, C. (2013). Tsunami resilience: Multi-level institutional arrangements, architectures and system of governance for disaster risk preparedness in Indonesia. *Environmental Science and Policy*, 29, 57–70.

United Nations. (2019). Goal 7: Affordable and clean energy. *Sustainable Development Goals*. Retrieved from www.un.org/sustainabledevelopment/energy/

United Nations Framework Convention on Climate Change. (2019). What is the Paris agreement? Retrieved from https://unfccc.int/process-and-meetings/the-paris-agreement/what-is-the-paris-agreement

Vrolijks, L. (1998). *Disaster resistant housing in Pacific Island countries: A compendium of safe low cost housing practices in Pacific Island countries*. New York: UN Department for Economic and Social Affairs.

Wikipedia. (2019). Biorock. Retrieved from https://en.wikipedia.org/wiki/Biorock

Zandvoort, M., Kooijmans, N., Kirshen, P., and van den Brink, A. (2019). Designing with pathways: A spatial design approach for adaptive and sustainble landscapes. *Sustainability*, 11(3), 565. doi:10.3390/su11030565

Environment Agency. (2017). Natural flood management – part of the nation's flood resilience. Retrieved from www.gov.uk/government/news/natural-flood-management-part-of-the-nations-flood-resilience.

Masters, B. (Director). (2017). 50 years ago, this was a wasteland. He changed everything. National Geographic YouTube channel. Retrieved from www.youtube.com/watch?v=ZSPkcpGmflE.

10 Sustained growth and development

While urban areas of the world appear to consistently be experiencing rapid expansion, rural areas tend to fluctuate, with periods of expansion and periods of decline. In the earlier chapters of this book we discussed some the reasons for this fluctuation. For example, the decline of small-holder agriculture is driving migration out of rural areas, while nature-based tourist attractions are enticing urbanites to purchase houses in a rural setting. In either case, affordable housing remains a great challenge, mainly because a fuller appreciation of what affordability means in a rural context is lacking. It is with this premise that the notion of affordability is revisited and freshly articulated using the core principles of sustainability: economic, social and environmental. Thus, the provision of affordable housing for rural communities will form an important part of their attainment of sustainable development goals.

The preceding chapters of this book reflected on the key issues affecting the chronic lack of affordable housing in a holistic manner. Here we bring these together to present a framework for developing and implementing affordable housing policies for rural communities. This framework incorporates a broad range of factors with an emphasis on the needs of the end-users. The framework is depicted in Figure 10.1.

One of the main arguments in this book is that housing affordability goes beyond the basic calculation of accommodation costs as a percentage of income to incorporate a range of other factors. An effective affordable housing solution must therefore firstly assess location specific demographic information. The community based demographic information needs to be carefully analysed with respect to the effectiveness of governance and other lateral support mechanisms which may be crucial to the development of the broader community.

As depicted in the figure, location specific demographic data comprises 16 core principles. These principles are briefly discussed below:

- Housing types and conditions: includes whether the house is vernacular or modern; whether the house is dilapidated or in a liveable condition; whether the house needs any repairs or refurbishments; a timeline of

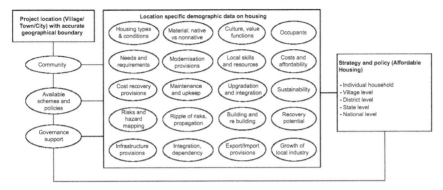

Figure 10.1 Framework for developing strategy and policy on affordable houses

how long the house is expected to remain operational; and the requirements of reconstruction or refurbishments including the extent of scope
and time

- Materials availability: native versus non-native: refers to the availability
  of local building materials; potential for use and re-use; longevity and
  maintenance requirements of local versus imported materials; and the
  dependence on imported materials from regional, national or even
  international markets
- Culture, value and functions of the house: incorporates the different
  levels of functionality a family may require from their home and its
  surrounding landscape. This includes a space for activities related to
  community culture, social values and religious rituals
- Occupants and their demographic profiles: refers to the number of
  household members, their ages and gender, their relationships to each
  other, their primary activities and socio-economic status of the individual families
- Needs and requirements: refers to space and other fundamental
  requirements of each member of the family as well as for the entire
  family and engagements within the community
- Modernisation provisions: refers to whether or not the existing house
  could benefit from modernisation, including an analysis of ways to perform these alterations in a cost and time effective manner
- Local skills and resources: incorporates a preference for utilising skills,
  resources and competencies within the family or the local community,
  avoiding reliance on external labour
- Costs and affordability: are sufficient funds available to each individual
  and the family as a whole to spend on accommodation costs? Whether
  there is sufficient income available after housing costs are covered to
  comfortably cover other basic essentials including cost of operation and
  maintenance of the house

- Cost recovery provisions: refers to the return on investment in both quantitative and qualitative terms, potential for families borrowing funds to make repayments with interest within the borrowed terms and conditions
- Maintenance and upkeep: refers to the potential, awareness and affordability of necessary home maintenance; access to required services and amenities; and keeping the property in the best possible condition to ensure it retains value over time
- Upgradation and integration: incorporates provisions to upgrade or integrate a house with other functional space that may be required over time to meet the emerging social or cultural functions
- Sustainability: involves technological solutions that incorporate environmental, economic and social principles into building or modernising a house
- Risks and hazard mapping: refers to assessing the potential risks and hazards, both natural and anthropogenic, that may disturb or destroy the house and its surrounding landscape
- Ripple of risks and propagation of risks: while a building may not suffer in a single hazard, the combined impact of many small hazards over time may lead to the accumulation of damage
- Building and rebuilding potential: the potential for building a new house or rebuilding a damaged house in the wake of a disaster
- Recovery potential: refers to the susceptibility to risks, coping mechanisms and resilience including time required for complete recovery in the event of a disaster
- Infrastructure provisions: the accessibility of the house's location to essential infrastructure
- Integration and dependency: the dependent relations between houses and infrastructure provisions for smooth functioning and adaptability to future infrastructure being planned in the vicinity
- Export/import provisions: refers to the growth and income potential benefitting from the improvised living conditions and self-empowerment
- Growth of local industry: the potential for contributing to the growth of local businesses; for generating employment opportunities for others; and progressively supporting the development of a healthy community.

An effective affordable housing strategy and its underlying policies should be adaptable to any level of abstraction from individual households to state or even international levels. Thus, rather than looking at the policies on the basis of "one size fits all", it needs be contextualised in relation to the considered location. The most effective affordable housing policies will be developed through systematic collection and processing of data from the target community.

The use of Information and Communication Technology (ICT) plays a crucial role in handling large demographic data. This has been demonstrated by one of the ongoing research projects within the Smart Villages Lab (SVL) at the University of Melbourne, Australia. The authors developed a unique ICT based Smart Data Platform and used it to collect socio-economic data from over 2000 households across 37 local villages on Majuli Island, Assam. The concept facilitates rapid assessment of community socio-economic conditions and needs by analysing data from the survey and integrating it with data from external sources. The integration capability of the external data assists in analysing how various intervention efforts on the selected community are aligned to the needs identified through the survey data. This resulted in a clearer understanding of the community's needs allowing more targeted and efficient governance outcomes. Details of the research was included in the authors' previous book (Doloi et al., 2019).

As the term "affordability" suggests, income generation is one of the key requirements for sustained growth and development of the community. However, ensuring a regular income stream for rural communities is challenging. The mainstream income generating opportunities found in urban areas simply do not exist in rural areas. There is significant potential for creating economic activities in rural areas, however, these are often unknown to locals living here. Identifying and harnessing these potentials by the local community is crucial to their development. In the previous book by the authors (Doloi et al., 2019), there is a chapter on "Income Generation" highlighting sustainable income generating opportunities suitable for rural areas, including businesses around agriculture, non-timber forest products and ecotourism.

This book has shown that while the provision of affordable housing incorporates a range of considerations, money still plays a big role in the construction of houses. The source of this necessary funding and its propagation through various stakeholders in affordable housing development schemes remains the most significant issue. As evidenced in this book, while the governments of developing countries tend to heavily subsidise housing, especially for rural communities, in some developed countries private investors are incentivised to develop affordable housing projects with a range of diverse business opportunities. A compromise between these two extremes, such as public-private partnerships, may be required for ensuring sustained growth and development especially in the rural community.

The heavy subsidisation of affordable housing programmes may not lead to the best outcomes for rural communities. Just having a roof over their head is not necessarily going to lead to long-term developmental improvements. Instead, affordable housing should be provided as part of a more holistic programme promoting expansion of rural economies. One option could be to require citizens to engage in activities that directly or indirectly enhance their opportunities for self-empowerment to qualify for housing assistance. Any housing assistance should not be seen as a long-term provision, but rather temporary assistance that effectively promotes citizens to self-sufficiency. These programmes should also not be targeted at individual households but aim to impact the broader community. For example, citizens could be required to actively participate in the construction of their house. Through this process they will learn construction skills, allowing them to both assist other families in their community in constructing a home, as well as passing on these construction skills to that household and so the developmental benefits should propagate through the community. Thus, the affordable housing programme needs to be derived as an enabler by integrating relevant conditions in local context. To be successful, such an approach would require a coordinated effort across government agencies, private organisations and members of the target community. The progress of such a scheme must be monitored to ensure ongoing success or identify points for future intervention.

References

Doloi, H., Green, R., and Donovan, S. (2019). *Planning, housing and infrastructure for Smart Villages.* Oxford: Routledge.

11 Epilogue

This book initiates fresh articulation of affordability in rural housing in the rapidly shifting balance between rural and urban. Departing from the conventional explanation of rural which is often discussed in contrast with urban, the book presents a comprehensive suite of indicators of affordability around social and cultural contexts that are unique to rural settings. A need for such a redefinition also arises from the sweeping changes that are occurring in rural areas such as continuing out-migration and demographic shifts, uneven distribution of wealth from emerging rural markets, land economics, changing aspirations, and cultural influences in construction. In short, the dynamics of houses and housing in rural areas are dramatically transforming making it necessary to revisit affordability. The book conceptualises affordability in rural housing along a spectrum that is interlaced with cultural and social values integral to rural livelihoods at both personal and community scales.

The book is developed around four intersecting themes: *explaining houses and housing in rural settings; exploring affordability in the context aspirations and vulnerability; rural development agendas involving housing and communities; and construction for resilience in rural communities.* For those residing in rural areas, constructing a house is a lifetime pursuit and hence, any work in progress with incremental quality improvement along with irregular resource flows is an important consideration in the affordability assessments. Such a practice is often nestled alongside broader socio-economic, cultural and political contours of rural community life and its incorporation into rural development policies. Affordable housing schemes also need to incorporate the influence of emerging technologies and digital illiteracies across rural communities.

Governments in emerging economies, where rural poverty is most acute, have attempted to improve rural development through numerous public schemes especially in the affordable housing contexts. However, due to their lack of understanding of the intrinsic factors around community and individual affordability, such as potential, aspirations, value and culture, these housing interventions have failed at multiple levels. Studies show that money allocated to public housing for rural communities is often not spent effectively due to the attempted imposition of urban development techniques on rural areas. Assessing affordability in rural communities and understanding the design and development of affordable housing that meets community expectations is a difficult task. Meeting such challenges requires its own techniques and approaches which need to be clearly consolidated and mapped in the rural contexts.

The authors, coming from both a research and professional background in housing, construction and infrastructure planning and development, have deep exposures and interests in rural life. Having conducted research and following some of the world practices in rural development, the authors have come to realise several things: firstly, that the challenges associated with measurement of affordability and the design of affordable housing for rural development needs completely new approaches and innovations not generally seen in cities, but nonetheless with potential application to city life; secondly, that the rural resilience, resourcefulness and community-spirit upon which such innovations draw are lacking in cities – but could greatly benefit cities, according to some authorities; thirdly, that for researchers and practitioners, doing rural work challenges many unexamined assumptions about how social and economic development "must" be done. Rural housing and rural development are mind-expanding and intellectually liberating, yet, the subjects are the least researched.

Building on the above knowledge gap, this book provides an overview of some of the little understood and sometimes counter-intuitive best practices on rural affordability and affordable housing that have emerged in uplifting rural communities in developing economies over the last 30 years. Drawing from the global literature and practice-based evidence across a complete spectrum of relevant disciplines, this book brings together a clear articulation of the innovative ideas around harnessing rural potential, and empowering rural communities with added affordability and progressive development in the context of housing and improved living standards.

Index

Page numbers in italics refer to figures. Page numbers in bold refer to tables.

Aboriginal Land Rights Act (1976), Australia 90
Abu-Ghazzeh, T. M. 24
Adamy, A. 163
Ademiluyi, Israel A. 91
adequacy, of housing 56–57; dilapidated houses 58–59, 91; health and safety 57–58; informal settlements 91; overcrowding 59–60; sense of community 58–59; urban *vs.* rural issues 58–59
aesthetics 3, 10, 29, 34, 60, 93, 161; *see also* culture
affordability 49, 68–69, 166; community 78–80; cooking facilities 76–78; defining 50, *51*; housing induced poverty 50, 57, 80; infrastructure 80–82; policy implications 82–83; rental accommodation *vs.* house ownership 54–55, *55*; solutions, legislative support for *61*; sustenance 73–76; transport 69–73
affordable housing 4, 48–49, 166; adequacy 56–60; availability 60–62; best practices 131–132; construction, process flow 88; definition of 49, *50*; engagement of citizens 169–170; and essential services employees 49; factors associated with 2–3, *2*; funding 63, 169; key dimensions of 6, *6*; location specific demographic information 166–168; negative image of 48; new terminology for 49; owner-occupied houses 51–53, 54–55, *55*; policy interventions 131–132; renting 54–56; role of government 62–65; strategy and policy, development of *167*

Africa: East Africa drought (2011–2012) 143; low-income families, inferior status of 20; mortgages in 53; quality of roads in 72
aged care facilities 21
agrarian communities 8, 15–17; access to water 81; and disasters 134; and earthquakes 137; resilience to flooding 160; tradition, loss of 7; vernacular architecture in 35, 41
agriculture 82; and farming subsidies 17; industrialisation of 1, 7, 15; intensive practices 75; monoculture 134, 160; produce loss, and droughts 143; and rainfall 142; slash and burn 144, 162; and sustenance 74; traditional agricultural practices 14; and transport 70
Al-Khaiat, Husain 39
Allen, B. L. 56
Alveano-Aguerrebere, I. 71–72
anaerobic digesters 77
Anna-Maria, Vissilia 34
aquaculture 75–75
arable land 68, 88; and housing development 10; and sustenance 74
Australia: average household size in 87; commuter communities in 9; cooperative housing schemes in 64; costs of housing in 48; heatwaves in 28; indigenous communities, land ownership of 90; lenders mortgage insurance 52; location of laundry facilities in 38; overcrowding in 59
Averda 102
Ayers, J. 160

Babu, B. V. 21
Babu, K. S. R. 154
Bach-Faig, A. 76
Badgirs 43
Bakar, A. H. A. 163
bamboo 95–96, 131, 153, 154–155, 158, 159
Bangladesh, flood resilience strategies in 160
Bański, Jerzy 8
basements, in vernacular architecture 31–32
bathrooms: and culture 38–39; in vernacular architecture 34, 38
Bhutan, alteration of buildings in 25
Bickford, Nate 14
biodiversity conservation 9, 88
biomass burning 28, 34, 80–81, 94; stoves 76
biorock 161–162
blackouts 28
Bodach, Susanne 29
Bramley, Glen 10
Brazil: community practice programme 64, 71; housing development schemes in 64
brown-field sites, agriculture in 74
build back better theory 138, 140, 145, 147, 150
businesses 2; in areas with high crime rates 79; and community 79; in commuter communities 11, 12; and green spaces 9; and mortgages 52; role, in housing provision 64; and technology 81–82; and tourism 13, 14; and urban migration 7; and vernacular architecture 40–41
ByFusion blockers 102

Canadian National Occupancy Standard 59
carbon monoxide poisoning, and low income housing 58
CARE Australia 141
Caribbean Islands, tourism communities in 14, 15
caste system, in India 20, 43, 89
ceiling height, in vernacular architecture 31
Chel, Arvind 94
chicken wire 100
children: childcare 64, 78–79; and community gardening 74, 75; of education 16, 42–43, 70, 72, 82; health of 57, 59, 60, 71, 75, 80; and

overcrowding 59, 60; role, in agrarian communities 15–16
China: average household size in 87; commuter communities in 10
cleaning habits 38
climate change: future projections related to 161; mitigation 150; and natural disasters 3, 143, 144
climatic responsiveness 3, 24, 27, 80; colder climates 28–29, 30–32; and culture, overlap of 44; energy consumption 27–28; extreme weather conditions and degree of tolerance 29–30, *30*; greenhouse gas emissions 27; and mud-bricks 94; thermal comfort 29–30; of vernacular architecture 28, 29–34, *31*, 41, 80, 130; warmer climates 28, 30, 31–34, 35, 38, 44
cluster housing 15
coastal communities, and tsunamis 139, 155–156
communities 78–80, 83, 131, 136, 166, 168, 170; commuter 8–12; and disaster rebuilding 140–141, 143, 145–146; and flood resistance 161; interaction, and access to transport 71; local, and disaster rebuilding 139; resilience of 151, *152*; resilient communal structure 157, 163; spirit 6, 11, 58–59, 68–69, 74, 140–141, 145; and traditional building styles 9
community centres/spaces 11, 44, 163
community gardening 74, 75, 164
community practice 64, 71
community scale projects 152
composite recycled plastic panels (CRPPs) 102
Conceptos Plasticos 102, 103
conflicts 135, 145–146; *see also* natural disasters
construction: of affordable houses, process flow 88; costs 3, 4, 87, 91, 92, 103, 117, 131; labour 80, 87, 91, 105, 136, 157; on-site, and weather conditions 103; skills, training 45, 105, 140; of vernacular architecture, knowledge/skills 36, 92; *see also* materials, building
construction industry 65; government schemes targeting 63–64; waste from 101
Cook, C. C. 71, 79
cooking facilities 76–78, 163
Cook Islands 13

Coombes, Mike 10, 60–61
cooperative housing schemes 64
corporate social responsibility (CSR) 14
corruption 62, 89, 156
Costa Rica 131, 153
courtyards, in vernacular architecture
33–34, 38, 39, 42, 44
Craigslist 81
critical regionalism 25–26, *26*
culture 3, 24, 36, 136; and alteration of
buildings 25; bathrooms and toilets 34,
38–39; cleaning habits 38; and climatic
responsiveness, overlap of 44; and
disaster rebuilding 139; and elderly care
20; extended families 39; and gendered
segregation 37–38; and historical events
7, 43; and housing layout 25; impact of
tourism on 13; and local food 77;
nature–culture–critical regionalism of
rural housing design 25–26, *26*; and
religion 36–37, 39; rituals, cooking/
eating as 76; of rural communities 6, 7;
traditions, loss of 7
cyclones 133, 139; Fani 141; Gita 141;
Pam 139–140; and resilience of
buildings 156–157; Winston 140–141
Cyprus, thermal conditions of school
buildings in 80

Dahshan, M. E. 143
Davis, Charles B. 68
Deakins, D. 81
demand response transport 73
dignity mortgages 52
Dili, A. S. 30, 33–34
disaster, definition of 133; *see also*
natural disasters
Dodo, Yakubu Aminu 29, 96
dormitory towns *see* manufactured
communities
Dorsey, R. W. 65
droughts 143–144, 162
Durand-Lasserve, Alain 91

early warning systems, tsunami 156
earthquakes 133, 135; Mexico 137–138;
Nepal 136–137, 151, 154; and
resilience of buildings 151, 153–155;
Turkey 135–136
East Africa drought (2011–2012) 143
eco-bricks 102
ecosystem services 9–10, 88
education: of children 16, 42–43, 70, 72, 82;
of community, about wildfires 162–163

Egypt: affordable housing schemes in 37;
vernacular architecture in 36, 92
ekra 129, 154, 155
elderly care, in rural communities 20,
21, 71
electric hobs 77
electricity: access to 1–2, 16, 81, 163;
transportation of 28; and vernacular
architecture 154
employment: in areas with high crime
rates 79; and chronic mobility 78;
and technology 18; in tourism
communities 14
end-users, consultation with/participation
of 44, 64–65, 83, 104
energy 109; consumption, by buildings
25, 27–28; efficiency 27, 80;
infrastructure 80–81
environment: benefits, of prefabrication
103; and building materials 95, 96;
and disaster rebuilding 143; eco-bricks
102; and fertilisers 17; and housing
developments 10; and prefabrication
103; and tourism 14; and vernacular
architecture 25, 29; and waste
management 101; *see also* climatic
responsiveness
evaporation, in vernacular architecture 34

families, extended 39
Fani (cyclone) 141
farm houses, renting of 13
farm tours 14
Fathy, Hasan 36, 92
Fereig, Sami M. 39
fertilisers 16–17, 75, 134
Fiji 140–141, 150
financial crisis (2007–2008) 52–53, 56
fires: and earthquakes 154; wildfires 134,
144, 162–163
fish farms 75–76
fishing 75, 139, 155
fixed rate mortgages 52
floating gardens 160
floating houses 159
floods 34, *35*, 41, 44, 142–143,
154–155; agricultural resilience to
160; and building materials 161–162;
floating houses 159; floor level,
raising of *158–159*; future climate
change projections 161; impacts on
buildings 158; and resilience of
buildings 158–162; soil absorbency
159–160

food: local, and tourism 77; poisoning, and low income housing 57; *see also* sustenance
Forbes, C. 136, 150, 151
forest dwelling communities 9–10
forest fires *see* wildfires
Forsyth, T. 160
Foruzanmehr, Ahmadreza 31, 32
fossil fuels 25, 28, 76, 77
foundations, building 34, *35*, 96–97, 100, 153, 154, 158, 159
fuels 27–28
Fuentes, José María 104

Gabriel, Michelle 92
Gallent, N. 72
Galloway, L. 81
Gautam, Avinash 29
Gediz earthquake (1970) 135–136
gendered segregation, in vernacular architecture 37–38
Gita (cyclone) 141
globalisation 7, 17, 24, 145
Good, Karen 11–12
Goswami, Monomoy 100
government 132, 171; demand side provision 106; interventions, affordable housing 62–65, 110–117; levels, differing responsibilities 82; supply side provision 105–106
grain storage 35, 42
greenhouse gas emissions 27, 77, 101, 133, 134
green spaces: in commuter communities 9–11; and flood resilience strategies 161; and tourism 82
Guerrero, L. 77
Gumtree 81

Habitat for Humanity 45, 64, 105, 140
Haigh, R. 145
Hamhaber, Johannes 29
hazardous materials, in low income housing 57
healthcare centres, and disaster resilience 163
healthcare facilities, access to 70–71
health problems: and affordable housing 57–58; and agriculture in brown-field sites 74; in colder climates 28–29; and indoor biomass/kerosene burning 76, 77; of overcrowding 59, 143; and toilets 39; in warmer climates 28
heatwaves 28

Henseler, M. 71
Heracleous, C. 80
Herman, K. 163
Hewson, B. 53
home ownership *see* owner-occupied houses
house size, and distance from urban centres 18, *18*
housing density 82, 161; in commuter communities 9, 10; and neighbourhood communication 18, *19*
housing for all initiative (India) 39, 89; *see also* Pradhan Mantri Adarsh Gram Yojana (PMAYG) houses
housing induced poverty 50, 57, 80
Hungary, indoor spaces in modern houses in 31
hurricanes *see* cyclones

Iceland, underground houses in 32
incineration 101
income generation 40–42, 145; and agriculture 74; and disaster rebuilding 143; eco-bricks 102; moving residents into single larger village 18; and Smart Village 1; and sustainability 169; and tourism 14
India: access to finance products in 53–54; aged care facilities 21; Assam Type houses 119–121, *119–121*, 123–125, *123–125*, 125, 129–130, *130*, 153–154, 158, *159*; biomass burning stoves in 76; building height restrictions in 43; caste system tradition in 20, 43, 89; combination of vernacular and modern architecture in 96, 97, 129, 155; commuter communities in 9–10; droughts in 144; floods in 142, 154–155, 158; forest dwelling communities in 9–10; gendered segregation in houses in 37; government affordable housing schemes in 62; impact of television 7; Kacha house 118, *118*, 121–123, *122–123*, 125, *129*; land ownership in 88–90, 91; manufactured communities in 18; modern houses in 34, *35*; PMAYG houses 110–117, *111–117*, **126–128**, 130–131; quality of roads in 72; seasonal rainfall in 142; self-employment in 41; self-made houses 117–125, *118–125*, **126–128**, 131; solar cell electricity programme in 81; sustenance in 74; thermal

comfort of vernacular architecture
29–30; vehicle ownership in 71;
vernacular architecture in 25, 29, 30,
33, 37, 37, 41, 125, 150, 153–154
Indian Ocean earthquake and tsunami
(2004) 138, 145, 155–156
indigenous communities, land rights
of 90
Indonesia: floating houses 159; forest
fires (1997–1998) 144; public transport
in 73; tsunami (2004) 155–156; vehicle
ownership in 71, 72
indoor ponds 43
indoor toilets 39
Indraganti, Madhavi 32
informal settlements 90–91
Information and Communication
Technology (ICT) 168–169
infrastructure 69; development, by
community 82–83; energy 80–81;
limitations of poor transport
infrastructure 69–71; quality of roads
to rural areas 72–73; technology
81–52; water and sanitation 81
insulation: and mud-bricks 94; in
vernacular architecture 31
interest only mortgages 52
interlocking bricks 154
internal nomadism 31, 32, 34
internet 2, 7, 16, 17, 24, 81–82
Iran: design restrictions in 43;
Zoroastrian community of 43, 44
Ireland: commuter communities in 11,
19; housing development policy in 82
Islamic culture 36–37, 38, 136

Jan Dhan Aadhaar Mobile (JAM) 54
Janssen, Jules 95, 153
Japan, vernacular architecture in 34,
40–41, 42
Jeffrey, Craig 89
Johnson, Kirk 7
Johnston, D. C. 71, 73
Jones, Roy 17
Joon, V. 76
juvenile detention 70

Kasapoglu, K. E. 154
Kaushik, H. 154
Kerala floods (2018) 142–143
kerosene stoves 76–77
keyworker housing 49
Kumar, D. S. 58
Kumar, J. M. 29, 91

Kusuma, Y. S. 21
Kutty, N. K. 50, 56

labour, construction 80, 87, 91, 105,
136, 157
land: arable 10, 68, 74, 88; availability
88–91, 132; informal settlements
90–91; legal fees associated with
buying 89, 91; ownership 88–90, 138;
rights, of indigenous communities 90;
tenure 4, 5, 89, 90, 91, 105, 138; *see
also* materials, building
landfills 101
landscapes: and earthquakes 142; and
floods 142; and thermal comfort 32
Lang, Werner 29
laundry rooms 38
Lautenschlager, L. 74, 75
Laxmi, V. 76
layout, housing 4, 25, 36, 38, 44, 63, 68;
see also location, housing
lenders mortgage insurance 52
Lerman, D. L. 59
Li, T. 56
lifecycle housing 49
lifestyle: and housing layout 36; of
indigenous communities 90; internal
nomadism 31, 32, 34; and internet/
television 7; rural 8, 17, 131; urban
34; *see also* culture
Light Gauge Steel (LGS) frames 96–97,
97–100, 103, 153
lighting, in vernacular architecture 34
living history, architecture as 7, 43–44
location, housing 3, 4, 68, 69; and
climatic responsiveness 30; and
cooking 78; demographic information
166–168; and earthquakes 154; and
mortgages 53; and rental assistance
programmes 79; and sustenance 74;
transport 69–73; and tsunamis 139,
155–156
Loza, Moises 68
Lucas, K. 70, 72, 73

Mace, A. 72
Maclennan, D. 49
maintenance *see* repairs/maintenance,
building
Maisonnave, H. 71
malaria 58
Malaysia: SMART village initiative in
17; vernacular architecture in 27
Malik, Jalaluddin Abdul 17

malls 11, 12
malnutrition 28; and droughts 143; and
    low income housing 57
Mani, Monto 30
manufactured communities 8, 17–18
materials, building 4, 87–88, 91–93;
    bamboo 95–96, 131, 153, 154–155,
    158, 159; biorock 161–162; and
    climate change 161; concrete 101, 153,
    154, 161; and construction costs 92;
    and earthquake collapse 153–154;
    ekra 129, 154, 155; and floods 161;
    mud-bricks 93–94; old and modern,
    combining 96–97, *97–100*, 129, 155;
    plastics 101–102; protocells 162;
    quality of 92; shortage, and disaster
    rebuilding 140
Mediterranean diet 76
Mexico: earthquakes (2017) 137–138;
    forest dwelling communities in 10;
    married life of women 20; modern
    housing design in 92–93; vehicle
    ownership in 72
Michael, A. 80
microgrid (electricity) 1, 16, 81
Millennium Development Goals 16, 42
mobile phones 16, 54, 73
modern architecture 3, 4, 24, 150;
    advantages of 34; bathrooms in 38–39;
    and climate 24, 29, 30; combining
    vernacular architecture with 25, 29,
    36, 96–97, *97–100*, 129, 155;
    cost-effective 100–105; and cyclones
    157; and earthquakes 137, 153; labour
    105; plastics 101–102; prefabrication
    103–104; preference for 34;
    repurposing existing buildings 104; *see
    also* vernacular architecture
monoculture 134, 160
Mora, R. 70
mortgages 51–52, 60, 79, 106; cooperative
    housing schemes 64; dignity 52; fixed
    rate 52; interest only 52; lenders
    mortgage insurance 52; and location of
    house 53; and middle class families 53;
    predatory lending 52; sub-prime 52–53;
    variable rate 52
Morton, L. W. 56, 58
mud-bricks 93–94
Mukhija, Vinit 92
multi-storey houses/buildings 39, 154
Mwase, N. 72
Myanmar, building height restrictions
    in 43

Naseer, M. A. 30, 33–34
natural disasters 3, 5, 133, *134*; cyclones
    139–141; droughts 143–144;
    earthquakes 135–138; floods 34, *35*,
    41, 44, 142–143, 154–155; rebuilding,
    consideration factors 143; tsunamis
    138–139; wildfires 144; *see also*
    resilience, of buildings
Nepal: earthquake (April 2015) 136–137,
    151, 154, 155; electrification in 16;
    tourism communities in 13;
    vernacular architecture in 41, 154
NevHouse 102, 103
Newport, S. J. 72
New Zealand, house prices in 49
Nguyen, M. T. 48
niche commodity markets 14, 17
niche tourism 77
Nigeria: building material costs in 91;
    combination of vernacular and
    modern architectures in 96
non-commercial fishing 75
non-governmental organisations (NGOs)
    54, 64, 91, 138, 146
Norway, small-holder farms in 14
Nuevas Escuales 16

orientation, vernacular buildings 30
outdoor cooking 77, 140
overcrowding 59–60, 143
owner-occupied houses 51–53, 54–55,
    55, 61, 62
ownership rights of land 88

Pam (cyclone) 139–140
Panagopoulos, T. 163
Papua New Guinea, land ownership of
    indigenous communities in 90
Pauli, Julia 20
pesticides 16–17, 75, 95, 134
Petcou, C. 83
Petrescu, D. 83
Pinard, M. I. 72
plastics 100–102
Portugal, rent control legislation in 55–56
Powers, William 68
Pradhan Mantri Adarsh Gram Yojana
    (PMAYG) houses 110–117, *111–117*,
    126–128, 130–131
Pradhan Mantri Awaas Yojana
    (PMAY) 109
predatory mortgage lending 52
prefabrication 103–104
protocells 162

public consultation *see* end-users, consultation with/participation of
public transport 72, 73, 82

rainfall 142
Raji, Bashiru A. 91
Ramamurthy, K. N. 64–65
Rapoport, Amos 39–40
Rashtriya Gramin Vikas Nidhi (RGVN) 54
Razak, Norizan Abudl 17
Reeder, W. J. 59
religion 36–37, 39, 139
rental accommodation 51, 54–56, 60, 61, 62; in commuter communities 8–9; *vs.* house ownership 55; impermanence 54; rent control 55–56; and repairs 54; in tourism communities 13
rental assistance programmes 56, 79
repairs/maintenance, building 54, 55, 58, 70, 105, 146, 151, 155, 157
repurposing, of buildings 104
resilience, of buildings 3, 5, 125, 143, 150–152; and cheap building materials 92; community scale projects 152; cyclones 140, 156–157; defining 163–164; earthquakes 136–137, 153–155; examination of surviving structures after disaster 151; floods 158–162; and housing designs 151; and natural disasters 150–151; tsunamis 138–139, 155–156; and vernacular architecture 35–36, 150–151, 153, 154–155; wildfires 162–163
Rijn, J. V. 72
roads to rural areas, quality of 72–73
Rönkkö, Emilia 21, 73
Rønningen, Katrina 14
roof top gardens 160
Royston, Lauren 91
Ruda, Gy 31
Rudofsky, Bernard 29
rural communities 1, 4, 6–8; affordable housing in 6; agrarian communities 15–17; atmosphere 10, 11, 12, 19; communication with global community 151; community spirit 6, 11, 58–59, 68–69, 74, 140–141, 145; commuter communities 8–12; functional requirements of rural houses 26, *26*, 36; manufactured communities 17–18; moving residents into single larger village 18–19; nature–culture–critical regionalism of

rural housing design 25–26, *26*; policy implications 19–21; rural charm, preservation of 11–12; tourism communities 12–15; and urbanites 8
Ryan, Robert L. 15

Saeed, Murad 17
safety, housing 3, 57–58, 102
Salesa, Jenny 105
Sanders, J. 81
San Francisco earthquake (1906) 154
sanitation: access to 81, 109; and droughts 143; in vernacular architecture 34
Saudi Arabia: defence-oriented design features in 43; modern architecture in 24
Sazinski, Richard J. 100
Sbarcea, M. 163
schooling 42–43; in agrarian communities 16; Nuevas Escuales 16; school to prison pipeline 70; and transport 70, 72, 78
Scotland, small-holder farms in 14
second homes 60, 61
seismic sea waves *see* tsunamis
self-employed individuals 135; and housing layout 25; mortgages 52; and transport 70; and vernacular architecture 40–41
self-made houses 117–125, *118–125*, **126–128**, 131
Sener, E. M. 80
shading, in vernacular architecture 32–33, *33*
Shastry, Vivek 30
short term leasing 56
Shucksmith, Mark 14
slash and burn agriculture 144, 162
small-holder farming 1, 7, 9, 14, 15, 17, 166
smart phones 16, 54
Smart Villages 1–3, *2*
Smart Villages Lab (SVL), University of Melbourne 169
Smith, C. 74, 75
socialisation: and community gardening 74; and community spaces 11; and culture 38; and transport 21, 71
social licence to operate (SLO) 14
socio-economic diversity, of communities 2, 59, 78–80
Solomons, M. 150
South Africa: impact of transport on healthcare access in 70; public

transport in 72, 73; upcycling of
plastic waste into eco-bricks in 102;
vehicle ownership in 71
Spain, tourism communities in 14, 15
squatter settlements 91
Sri Lanka 145
stilt houses 34, *35*, 41, 44, 97, *100*, 150,
154–155, 158, *158*
storms *see* cyclones
stress, and low income housing 57
sub-prime mortgages 52–53
subsidies 63, 169
sustainability 10, 28, 76, 166; and Assam
Type houses 125, 129; and cooking
practices 77; and income generation
169; and locally sourced materials
129, 155
Sustainable Development Goals 16, 39,
42, 57, 69, 71, 76, 81, 109
sustenance 68, 73–76, 137, 160

technology 81–82; and access to finance
products 54; Information and
Communication Technology 168–169;
and manufactured communities 18;
and transport 73
television 7, 16, 24
Tenorio, Rosangela 30
tenure, land 4, 5, 89, 90, 91, 105, 138
Tewdwr-Jones, M. 72
thatched roofs 94, 130, 157
thermal comfort 80; and degree of
tolerance 29–30, *30*; and low income
housing 57; of vernacular architecture
29–30, 32
Tiwari, G. N. 94
toilets 38–39
Tonga: cyclone Gita 141; schooling in
42–43
Tonts, Matthew 17
tourism 166; in agrarian communities
17; communities 8, 12–15; in
commuter communities 11; and
fishing 75; and green spaces 82; and
local traditional food 77; natural
beauty, and housing development 14,
15; and vernacular architecture 13, 34
traditional agricultural practices 14
transport 68; costs, of building materials
91–92; and home cooking 78;
limitations of poor transport
infrastructure 69–71; modes, access to
71–73; public transport 72, 73, 82;
and sustenance 74

trees, use in vernacular architecture
32–33
Trinidad and Tobago 82–83
Troell, M. 75
Trowbridge, A. V. 17, 18, 21
tsunamis 133, 138–139, 145; and
resilience of buildings 155–156;
warning 156
Turkey, Gediz earthquake (1970) 135–136
typhoons *see* cyclones

United Kingdom: availability of
properties 60; commuter communities
in 10; housing development schemes
62; location of laundry facilities in 38;
rural areas, sense of community in 59;
tourism communities in 15
United Nations 16, 42, 57, 69, 71, 74, 76,
81, 90, 109, 138
United Nations Education, Scientific and
Cultural Organisation (UNESCO) 76
United States: adequacy of housing in
rural areas in 58–59; affordable
housing in 48; average household size
in 87; communal gardening in 74, 75;
commuter communities in 10–12;
flood resilience strategies in 160–161;
mortgages in 52, 53; prefabrication
industry 104; preservation of rural
charm in 11–12; rental assistance
programmes in 56, 79; San Francisco
earthquake (1960) 154; vernacular
architecture in 33
urban migration 7–8, 15, 20–21, 24,
71, 166

Vanuatu 139–140, 150
Varghese, T. Zacharia 30, 33–34
variable rate mortgages 52
Varma, G. R. 21
Vastu Shastra 38
Vellinga, Marcel 31, 32
vernacular architecture 3, 4–5, 24, 25, 92,
125; bamboo 95–96, 131, 153, 154–155,
158, 159; basements 31–32; ceiling
height 31; climatic responsive features
of 28, 29–34, 41, 80, 130; in colder
climates 28, 30–32, 77; combining with
modern architecture 25, 29, 36, 96–97,
97–100, 129, 155; construction,
knowledge and skills 36, 92; courtyards
33–34, 38, 39, 42, 44; cultural features
of 37, 38–39, 40, 42; and cyclones 157;
disadvantages of 34–35, 93; and

earthquakes 136–137, 153, 154–155; environmental sustainability of 25; evaporation 34; and extended families 39; and fires 154; floating houses 159; and floods 158; gendered segregation in 37–38; in hot/humid climate, key characteristics 30, *31*; and income generation 41; insulation 31; local perception of 92–93; longevity of 35–36; mud-bricks 93–94; orientation 30; outdoor cooking 77; rejection, reasons for 34; research into 27; and resilience of buildings 35–36, 150–151, 153, 154–155; safety of 35; shading 32–33, *33*; thermal comfort of 29–30; and tourism 13, 34; in warmer climates 28, 30, *31*–34, 35, 38, 44, 77; windows 33; *see also* modern architecture
volunteers, disaster building 146

Walker, K. L. M. 144
walls, building 93, 94, 97, 100, 129, 130, 153
WasteAid UK 102
waste management 75, 100–101, 103

water, access to 81, 109, 164
Watkins, David 10
Watson, Georgia Butina 25, 40
weather: conditions, and on-site construction 103; reports, and television access 16; *see also* climatic responsiveness
Wesołowska, Monika 8
wildfires 134, 144; education about 162–163; and resilience of buildings 162–163
Williams, R. 49
windows, in vernacular architecture 33
Winston (cyclone) 140–141
women: abuse, after marriage 20–21; banking access, in India 54; and post-conflict rebuilding 145; role, in agrarian communities 15
workforce housing 49
World Habitat 91

Zandvoort, M. 160
Zetter, Roger 25, 40
Ziebarth, A. 58
Zimbabwe, tourism communities in 14

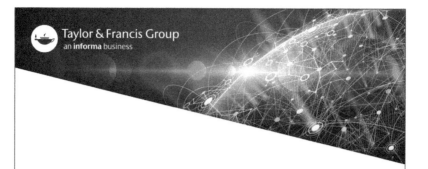